Ecotourism Horizons

This book invites readers on a captivating journey through the complexities of ecotourism around the world. *Ecotourism Horizons: From Community Empowerment to Global Perspectives* offers a comprehensive exploration of the diverse landscapes and issues within the realm of ecotourism, divided into two insightful parts.

Part I unveils community-based ecotourism's transformative potential, from homestays in the Himalayas to empowerment in Malaysian Borneo and Cambodia. Gender dynamics are explored alongside community empowerment in this enlightening section, shedding light on the nuanced relationships between tourism, society, and environment. Part II explores global insights on ecotourism, spanning destinations like China, India, Iran, and sub-Saharan Africa. From outbound Chinese tourists' experiences to challenges faced by ecolodges, diverse perspectives illuminate ecotourism's multifaceted landscape, addressing local initiatives and broader issues.

A must-read for enthusiasts and scholars alike, *Ecotourism Horizons: From Community Empowerment to Global Perspectives* offers invaluable insights into the complexities and potentials of ecotourism worldwide. Discover how communities, tourists, and environments interact in this dynamic and evolving field.

The chapters in this book were originally published in *Tourism Recreation Research.*

Tzung-Cheng (TC) Huan is Professor at National Chiayi University, Taiwan. He is also the Editor-in-Chief of *Tourism Recreation Research* and a member of the UNWTO (United Nations World Tourism Organization) Panel of Tourism Experts. He has been honoured by the 2011–2012 Outstanding Reviewer Award from *Cornell Hospitality Quarterly*; the 2013, 2017, and 2018 Emerald Literati Awards; and the 2018 Publons Peer Review Award.

Aliana Man Wai Leong is Associate Vice-President for International Affairs and Dean, Professor, and Doctoral Supervisor in the School of Liberal Arts, Macau University of Science and Technology (MUST), China. She is also the former Dean of the Faculty of Hospitality and Tourism Management, MUST. She has been honoured as one of the "Top Ten Most Influential Teachers in China's Tourism and Hospitality Education" and one of the "Top Ten Most Influential Teachers in National Tourism Institutions."

Tao Ma is Professor at the School of Management and the Dean of the Business School at Harbin Institute of Technology, China. His research interests include regional industrial planning and the economic analysis of resources and energy. He is recognized as a "Youth Talent" under the National Cultural Masters and Four Batches Talent Project by the Publicity Department of the Communist Party of China and serves as the Chief Expert for major projects funded by the National Social Science Fund of China.

Ecotourism Horizons

From Community Empowerment to Global Perspectives

Edited by
Tzung-Cheng (TC) Huan, Aliana Man Wai Leong and Tao Ma

Routledge
Taylor & Francis Group
LONDON AND NEW YORK

First published 2025
by Routledge
4 Park Square, Milton Park, Abingdon, Oxon OX14 4RN

and by Routledge
605 Third Avenue, New York, NY 10158

Routledge is an imprint of the Taylor & Francis Group, an informa business

British Library Cataloguing in Publication Data
A catalogue record for this book is available from the British Library

ISBN13: 978-1-032-86273-6 (hbk)
ISBN13: 978-1-032-86276-7 (pbk)
ISBN13: 978-1-003-52214-0 (ebk)

DOI: 10.4324/9781003522140

Typeset in Myriad Pro
by Newgen Publishing UK

Publisher's Note
The material in this volume has been reproduced in facsimile. This means we can retain the original pagination to facilitate easy and correct citation of the original essays. It also explains the variety of typefaces, page layouts and numbering.

The publisher accepts responsibility for any inconsistencies that may have arisen during the conversion of this book from journal articles to book chapters, namely the inclusion of journal terminology.

Disclaimer
Every effort has been made to contact copyright holders for their permission to reprint material in this book. The publishers would be grateful to hear from any copyright holder who is not here acknowledged and will undertake to rectify any errors or omissions in future editions of this book.

Contents

Citation Information

The following chapters were originally published in various issues and volumes of *Tourism Recreation Research*. When citing this material, please use the original page numbering for each article, as follows:

Chapter 1

Homestays' contribution to community-based ecotourism in the Himalayan region of India
Priya Bhalla, Alexandra Coghlan and Prodyut Bhattacharya
Tourism Recreation Research, volume 41, issue 2 (2016), pp. 213–228

Chapter 2

Community-based ecotourism managing to fuel community empowerment? An evidence from Malaysian Borneo
Velan Kunjuraman
Tourism Recreation Research, volume 47, issue 4 (2022), pp. 384–399

Chapter 3

Community-based ecotourism and the transformative learning of homestay hosts in Cambodia
Vicheth Sen and Pierre Walter
Tourism Recreation Research, volume 45, issue 3 (2020), pp. 323–336

Chapter 4

Gender Analysis in Community-based Ecotourism
Pierre Walter
Tourism Recreation Research, volume 36, issue 2 (2011), pp. 159–168

Chapter 5

Experiences and value perceptions of an ecotourism trip – an empirical study of outbound Chinese tourists
Ting (Tina) Li, Fang Liu and Geoffrey N. Soutar
Tourism Recreation Research, volume 46, issue 3 (2021), pp. 333–344

Chapter 6

Rapid ecotourism development and challenges: the case of ecolodge in Iran
Hamid Reza Irani, Mehdi Golverdi, Raymond Rastegar and Soheil Rezaei
Tourism Recreation Research
DOI: https://doi.org/10.1080/02508281.2022.2086417

Chapter 7

Ecotourism in Yancheng Wetland, China
Feifei Xu and Geoffrey Wall
Tourism Recreation Research, volume 32, issue 1 (2007), pp. 11–19

Chapter 8

Ecotourism in Gangotri Region of the Garhwal Himalayas
Gitanjali Chaturvedi
Tourism Recreation Research, volume 27, issue 3 (2002), pp. 41–51

Chapter 9

Globalization, Ecotourism and the Promise of Development in Sub-Saharan Africa
Andrew Lepp
Tourism Recreation Research, volume 27, issue 1 (2002), pp. 61–68

Chapter 10

Local Initiatives to Localize Ecotourism: An Exploratory Study in Kerwa, Van Vihar National Park Catchments
A. K. Bhattacharya, Suchitra Banerjee and Vandana Saksena
Tourism Recreation Research, volume 28, issue 1 (2003), pp. 97–102

Chapter 11

Ecotourism Issues and Options for New Zealand
Ross Cullen, Kay Booth and Ken Hughey
Tourism Recreation Research, volume 28, issue 2 (2003), pp. 89–92

Chapter 12

Ecotourism in the Australian and New Zealand Sub-Antarctic Islands
C. Michael Hall
Tourism Recreation Research, volume 18, issue 2 (1993), pp. 13–21

Notes on Contributors

Suchitra Banerjee is Professor at Institute for Excellence in Higher Education (IEHE) Bhopal, India.

Priya Bhalla is Lead Auditor and Technical Reviewer at SCS Global Services. She completed her Ph.D. thesis on ecotourism impact assessment from Guru Gobind Singh Indraprastha University, Delhi. She has a special interest in nature-based tourism management and currently is focusing on ecotourism practices at Binsar Wildlife Sanctuary, Uttarakhand, India.

A. K. Bhattacharya is Chairman of Skills Art & Beyond, and former Associate Professor at Indian Institute of Forest Management, Bhopal, India.

Prodyut Bhattacharya is Professor in the University School of Environment Management, Guru Gobind Singh Indraprastha University, Delhi. His research interests pertain to the broad natural resource management issues and livelihood linkages; specifically in the areas of biodiversity conservation, forest ecology, sustainable NTFP management, effect of climate change, urban forestry, forest governance, certification, and sustainable forest management.

Kay Booth is Executive Director at the Institute of Public Administration New Zealand.

Gitanjali Chaturvedi is a former research scholar in tourism, Centre for Political Studies, Jawaharlal Nehru University, Delhi, India.

Alexandra Coghlan is Associate Professor in the Department of Tourism, Sport and Hotel Management, Griffith University, Australia. Her research focuses on tourist behaviour, with a particular focus on pro-environmental and pro-social behaviour, and the links between active travel, well-being, and pro-environmental behaviour.

Ross Cullen is Emeritus Professor at Lincoln University, Canterbury, New Zealand.

Mehdi Golverdi is Assistant Professor of Public Administration at Department of Humanities and Social Sciences, Ardakan University, Ardakan, Iran. He gained Ph.D. and master degrees from the University of Tehran in public administration. He is interested in behavioural public policy, organizational behaviour, and qualitative research.

C. Michael Hall is currently Professor in Marketing and Tourism in the Department of Management, Marketing & Tourism, School of Business, University of Canterbury, Christchurch, New Zealand, having joined the department in 2007. He is also currently Visiting Professor and Docent in Geography, University of Oulu, Finland; Visiting Professor in the Tourism Studies programme, School of Business & Economics, Linneaus University, Kalmar, Sweden; Guest Professor, Department of Service Management at Lund Helsingborg campus, Sweden; Visiting Professor, CRIT (Centre for Research and Innovation in Tourism), Taylors University, Kuala Lumpur, Malaysia; and Eminent Scholar in the College of Hospitality and Tourism Management, Kyung Hee University, Seoul, South Korea. He is also a frequent visitor to the geography department of Umeå University, also in Sweden.

Tzung-Cheng (TC) Huan is Professor at National Chiayi University, Taiwan. He is also the Editor-in-Chief of *Tourism Recreation Research* and a member of the UNWTO (United Nations World Tourism Organization) Panel of Tourism Experts. He has been honoured by the 2011–2012 Outstanding Reviewer Award from *Cornell Hospitality Quarterly*; the 2013, 2017, and 2018 Emerald Literati Awards; and the 2018 Publons Peer Review Award.

Ken Hughey is Professor of Environmental Management at Lincoln University, Christchurch, New Zealand.

Hamid Reza Irani is Assistant Professor of Entrepreneurship and Marketing at Department of Business Management, College of Farabi, University of Tehran, Iran. He is the Director-in-Charge of *Iranian Journal of Management Studies* and an editorial board member of *International Journal of Organizational Leadership*. He is interested in digital marketing, entrepreneurship, tourism, and marketing channels. His research appeared in refereed academic journals such as *Advances in Hospitality and Tourism Research*, *Journal of Entrepreneurship & Organization Management*, and *Encyclopedia of Tourism Management and Marketing*.

Velan Kunjuraman is currently affiliated with Universiti Malaysia Kelantan (UMK), Malaysia, as a tourism lecturer. He has received his Ph.D. at Universiti Malaysia Sabah, East Malaysia. Dr. Velan Kunjuraman has been teaching a variety of subjects in the tourism field including tourism planning and development, tourism product development, tourism marketing and heritage, and cultural tourism. His research interests are in sociology of ecotourism and community development, seaweed cultivation and homestay programme, community-based ecotourism, tourism planning, and tourism impacts.

Aliana Man Wai Leong is Associate Vice-President for International Affairs and Dean, Professor, and Doctoral Supervisor in the School of Liberal Arts, Macau University of Science and Technology (MUST), China. She is also the former Dean of the Faculty of Hospitality and Tourism Management, MUST. She has been honoured as one of the "Top Ten Most Influential Teachers in China's Tourism and Hospitality Education" and one of the "Top Ten Most Influential Teachers in National Tourism Institutions."

Andrew Lepp is Professor of Recreation, Park, and Tourism Management at the Kent State University, Ohio, United States.

Ting (Tina) Li is Associate Professor at Hainan University-Arizona State University Joint International Tourism College, China.

Fang Liu holds a Ph.D. in marketing and is Associate Professor in Marketing at the Business School of the University of Western Australia (UWA). Dr. Liu's research, teaching, and consulting areas centre around consumer/tourist experience, advertising, marketing communication, branding, consumer psychology, international trade, product/service innovation and development, and cross-cultural consumer behaviour. She has a wide range of publications, including dozens of refereed journal articles and book chapters.

Tao Ma is Professor at the School of Management and the Dean of the Business School at Harbin Institute of Technology, China. His research interests include regional industrial planning and the economic analysis of resources and energy. He is recognized as a "Youth Talent" under the National Cultural Masters and Four Batches Talent Project by the Publicity Department of the Communist Party of China and serves as the Chief Expert for major projects funded by the National Social Science Fund of China.

Raymond Rastegar holds a Ph.D. in tourism management and is a lecturer in the Department of Tourism, Sport and Hotel Management, Griffith University, Australia. His research focuses on sustainable tourism development in developing countries with specific interest in local community livelihood, participation, empowerment, and environmental conservation. Recent research projects have investigated the impacts of tourism at local level and delivered new insights into the tourism phenomenon to advocate a more just and sustainable tourism future.

Soheil Rezaei has an MBA degree from the University of Tehran. His research focuses on tourism and ecotourism development in Iran.

Vandana Saksena is former Researcher at the Centre for Environmental Sciences, Jiwaji University, Gwalior, India.

Vicheth Sen is a Ph.D. candidate in the Department of Educational Studies, University of British Columbia, Canada.

Geoffrey N. Soutar is Winthrop Professor at the University of Western Australia (UWA). An economics graduate from UWA, Dr. Soutar undertook doctoral training at Cornell University before returning to teach at the UWA from 1973 to 1986. He was Foundation Professor of Management at Curtin University of Technology, Australia, from 1986 to 1994 and Executive Dean of the Faculty of Business and Public Management at Edith Cowan University, Australia, from 1994 until 1999. He was also Director of the Graduate School of Management at UWA from 2000 until 2007 and Head of the Marketing Discipline Group at UWA from 2008 to 2016. Prof. Soutar has been a consultant to a large number of private and public sector organisations in Australia and internationally and has been active in research across a wide area, publishing more than 200 research papers in journals and in book chapters, as well as a number of research monographs, across a wide range of management and marketing areas and presenting more than 300 papers at seminars and conferences. His present research interests include cross-cultural decision-making, new product and service development and the marketing of services, especially educational and tourism services. He has a particular interest in service quality and its impact on organisational success, from which evolved a long-term study of consumption value and its impact on people's willingness to buy and their subsequent satisfaction or dissatisfaction.

Geoffrey Wall is Distinguished Professor Emeritus of Geography in the Faculty of Environmental Studies, University of Waterloo, Ontario, Canada.

Pierre Walter is Professor in the Department of Educational Studies, University of British Columbia, Vancouver, Canada. His research interests include community-based ecotourism, learning in the climate justice movement, and decolonising living history museums.

Feifei Xu is Lecturer in Tourism, Leisure and Events, School of Services Management, Bournemouth University, United Kingdom.

Introduction

Tzung-Cheng (TC) Huan, Aliana Man Wai Leong and Tao Ma

In an era where sustainable development is increasingly critical, ecotourism stands out as a promising approach to harmonize human interaction with the environment. *Ecotourism Horizons: From Community Empowerment to Global Perspectives* embarks on an enlightening journey through the complex world of ecotourism, offering a rich tapestry of insights that span the globe. This book delves into the multifaceted realm of ecotourism, exploring its potential to foster community empowerment and address global challenges.

The first part of this volume, "Global insights on community-based ecotourism", illuminates how community-focused initiatives in ecotourism can drive significant social and environmental change. Here, we explore the transformative power of ecotourism through the lens of local experiences and empowerment. The chapters offer a deep dive into the impact of community-based ecotourism from the Himalayan region of India to Malaysian Borneo and Cambodia.

Priya Bhalla, Alexandra Coghlan, and Prodyut Bhattacharya's examination of homestays in the Himalayas reveals how these initiatives contribute not only to economic benefits but also to cultural preservation and community cohesion. The insights provided underscore the dual role of ecotourism in enhancing both the livelihoods of local communities and the cultural heritage of the region.

Velan Kunjuraman's study on community-based ecotourism in Malaysian Borneo examines how these projects can fuel community empowerment. By analyzing local experiences and the broader impacts of ecotourism, this chapter highlights the transformative potential of ecotourism when managed effectively.

In Cambodia, Vicheth Sen and Pierre Walter explore how homestay hosts undergo transformative learning through their involvement in community-based ecotourism. This chapter offers a compelling look at the personal and communal growth that can result from ecotourism initiatives, reflecting on the broader implications for sustainability and development.

Pierre Walter's contribution on gender dynamics within community-based ecotourism provides a critical analysis of how gender roles and relations are influenced by and contribute to ecotourism efforts. This perspective is crucial for understanding the intersection of social issues with ecotourism practices and outcomes.

Transitioning to Part II, "Global insights on ecotourism", the book shifts focus to a broader view of ecotourism's impact and challenges worldwide. This section explores diverse destinations and perspectives, reflecting the global nature of ecotourism and its varied manifestations.

Ting (Tina) Li, Fang Liu, and Geoffrey N. Soutar's empirical study on outbound Chinese tourists' experiences and value perceptions offers insights into the growing influence of Chinese tourists on the global ecotourism market. Their findings highlight the evolving preferences and expectations of a significant segment of global travelers.

The chapter by Hamid Reza Irani, Mehdi Golverdi, Raymond Rastegar, and Soheil Rezaei on rapid ecotourism development and challenges in Iran provides a detailed analysis of the issues faced by ecolodges. This study sheds light on the complexities of balancing growth with sustainability in a rapidly changing context.

Feifei Xu and Geoffrey Wall's exploration of ecotourism in Yancheng Wetland, China, and Gitanjali Chaturvedi's study on the Gangotri region of the Garhwal Himalayas offer case studies that highlight the diverse environmental and cultural contexts in which ecotourism operates. These chapters provide valuable lessons on the local adaptations and innovations within the field.

Andrew Lepp's discussion on globalization, ecotourism, and development in sub-Saharan Africa addresses the promise and pitfalls of ecotourism as a development tool. This chapter critically examines the impact of ecotourism on local communities and environments in a region facing significant developmental challenges.

A. K. Bhattacharya, Suchitra Banerjee, and Vandana Saksena's exploratory study on local initiatives in Kerwa, Van Vihar National Park catchments, offers a closer look at grassroots efforts to localize ecotourism. Their research emphasizes the importance of local engagement and tailored approaches in achieving sustainable outcomes.

Ross Cullen, Kay Booth, and Ken Hughey's analysis of ecotourism issues and options for New Zealand provides a comprehensive overview of the unique challenges and opportunities within this context. Similarly, C. Michael Hall's exploration of ecotourism in the Australian and New Zealand sub-Antarctic islands underscores the distinct environmental considerations in ecotourism practices.

Ecotourism Horizons: From Community Empowerment to Global Perspectives serves as a crucial resource for understanding the complex interactions between tourism, community, and environment. It offers a wealth of knowledge for researchers, practitioners, and anyone passionate about sustainable travel and responsible tourism. By bringing together diverse perspectives and experiences, this book illuminates the evolving landscape of ecotourism and its potential to contribute to a more sustainable future.

As you delve into the chapters of this book, we invite you to reflect on the intricate balance between human activities and environmental stewardship. The insights presented here aim to inspire and inform, fostering a deeper understanding of how ecotourism can both empower communities and address global challenges.

Part I

Global insights on community-based ecotourism

Homestays' contribution to community-based ecotourism in the Himalayan region of India

Priya Bhalla, Alexandra Coghlan and Prodyut Bhattacharya

ABSTRACT

This article investigates how villagers' participation in the homestay programme can influence attitudes and behaviours related to ecotourism objectives within a wildlife sanctuary. Initially, it provides a historical context of the development of the homestay programme within the Binsar Wildlife Sanctuary, situated in Kumaon Hills of the Indian Himalayan region using a case-study approach. Based on interviews with each household head conducted within the Sanctuary, the paper explores the links between villagers' homestay involvement and, attitudes and behaviours related to the Sanctuary's ecotourism objectives. The findings suggest that contextual variable such as occupation significantly influences villagers' attitudes towards the homestay programme while human–wildlife interactions additionally influence the villagers' attitudes towards ecotourism development. Furthermore, positive attitudes towards homestays have been manifested as positive ecotourism-directed behaviours resulting in villagers' engagement in public–private partnerships, their involvement in tourism-related cultural programmes and willingness to contribute towards nature interpretation activities to support ecotourism objectives in the Sanctuary. While the homestay programme has encouraged local guardianship behaviour, opportunities to improve and expand conservation efforts with the help of communities in the region could be further developed, and require greater cooperation from concerned stakeholders in both public and private sectors.

Introduction

While primarily a nature-based tourism product, the definitions and practices of ecotourism often include a social dimension (Fennell, 2015). The direct or indirect involvement of local people in community-based ecotourism (CBE) is considered desirable, as locals can play an important role in achieving the goals of ecotourism to protect the natural environment (Regmi & Walter, 2016; Reimer & Walter, 2013; Stone, 2015). This study explores the effects of including a local CBE product, homestays, within an ecotourism destination of Binsar Wildlife Sanctuary (BWLS) in the Indian Himalayan Region of Uttarakhand in Northern India (Figure 1).

The homestay tourism product development in BWLS represents an interesting case of local resident-based community involvement within an area which is considered of high conservation value (Table 1). The tourism initiative aimed at encouraging employment and revenue for villagers living within the Sanctuary, providing an alternative to extractive or consumptive activities (Drumm & Moore, 2005) and reduce outward migration (Chaturvedi, 2002). Our study explores whether the inclusion of homestays within the Sanctuary's ecotourism strategy stimulates local guardianship of natural resources and provides support for ecotourism objectives through changes in villagers' attitudes and behaviours towards conservation and ecotourism initiatives in the Sanctuary.

Ecotourism and local guardianship of natural resources

It is well understood now, that support of local communities is indispensable for the success of any protected area (PA) management (Drumm & Moore, 2005; Gurung, 1995; Lindberg & Hawkins, 1999; Mehta & Heinen, 2001; Rastogi, Badola, Hussain, & Hickey, 2010; Sinha, Qureshi, Uniyal, & Sen, 2012; Wells & Brandon, 1993). Researchers suggest that by creating economic incentives for the local villagers, ecotourism can encourage local guardianship of biological resources and assist in improving surrounding environmental conditions (Bookbinder, Dinerstein, Rijal, Cauley, & Rajouria, 1998; Kala, 2013). By creating direct linkages between

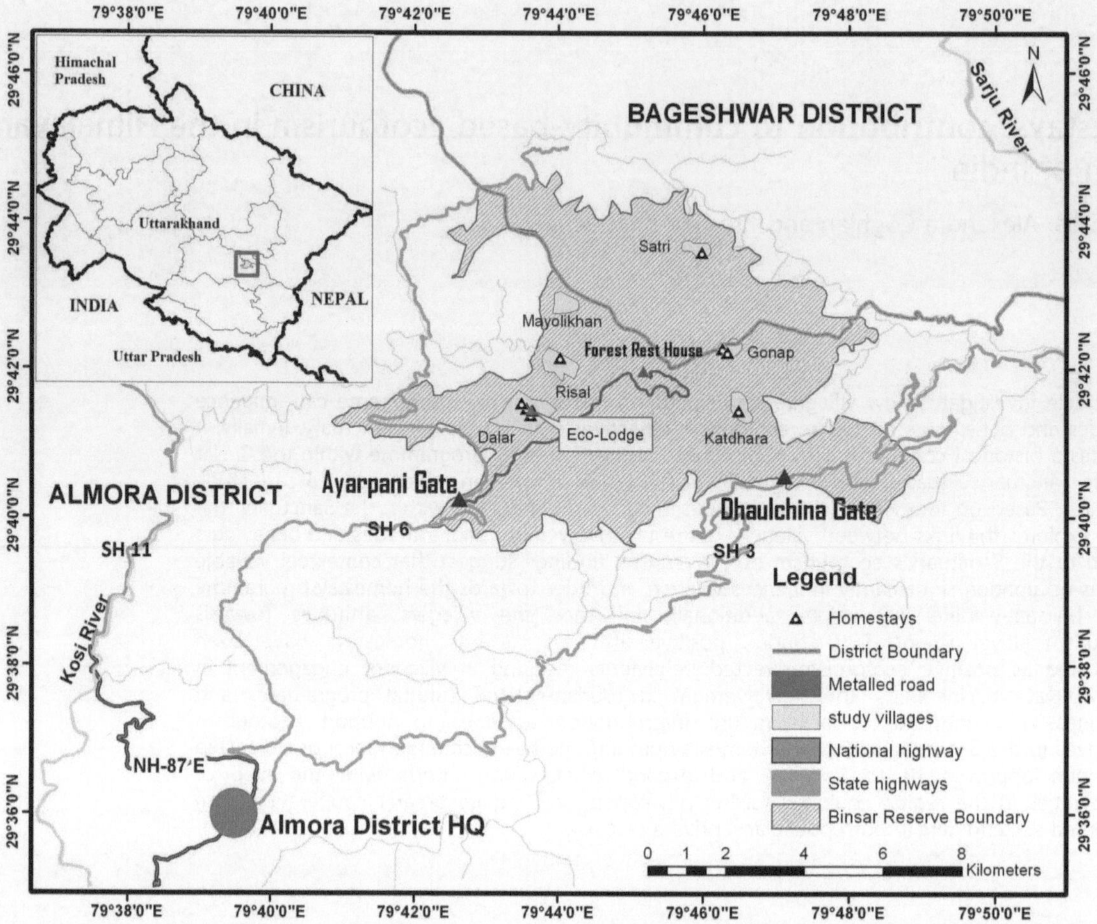

Figure 1. Location map of BWLS (prepared by P. Bhalla, with permission of G. Areendran, Head, IGCMC, WWF-India).

Table 1. Characteristics of Binsar protected area.

Characteristic	BWLS
PA category	IUCN category IV (habitat/species management areas)
Size	47 km^2
Location	29°39′N–29°44′N, 79°41′E–79°49′E
Established	1988
Tourists/year	18,809 (April 2014 – March 2015)
Vegetation	Himalayan moist temperate forest, Oak (*Querecus* sp.) and Pine (*Pinus* sp.)
Key species	Fauna (Leopard, Ghoral, Serow, Jackal, Barking deer)
Designation	Important Bird Area, A3 category of Biome 08 (Birdlife International, 2016)
Human settlements inside PA	Five villages
Accommodations inside PA	17

people and conservation, ecotourism can help in changing the attitudes of local people towards conservation of biodiversity and reduce their dependence on natural resources like timber, non-timber forest products (NTFP), fuelwood, fodder, thatching materials and other forest products (Das & Chatterjee, 2015; Nyaupane & Poudel, 2011; Regmi & Walter, 2016). But considering the non-homogeneous nature of participating communities and their diverse interests in conservation and development issues, researchers highlight the importance of studying target communities values, interests and attitudes (Brandon, 2001; Brandon & Margoluis, 1996; Heinen, 1996; Okazaki, 2008). Cobbinah's (2015) study in Ghana, Africa, revealed that locals' positive attitudes towards conservation are influenced by accruing socio-economic benefits in terms of employment, income and involvement in natural resource management and suggests that raising their knowledge of the environmental benefits can increase local support for conservation. This is also evident from the study by Bookbinder et al. (1998) in the Royal Chitwan National Park, Nepal where a community-based microenterprise approach helped in strengthening local stewardship towards biodiversity conservation along with directing a substantial amount of revenue to local development.

Furthermore, Zhang and Lei (2012) explored factors contributing to residents' participation intention in ecotourism management and proposed a structural relationship between their participation intention, environmental knowledge, attitudes towards ecotourism and the appeal of tourists landscapes. They found that residents' environmental knowledge positively affects attitudes towards ecotourism, which in turn directly and indirectly determine the intention to participate in ecotourism through their individual landscape affinity. They further used 'ecological monitoring', 'training programmes', 'encouraging conservation participation' and 'discussion-meetings' as items to measure attitudes of residents towards ecotourism. Various other variables and attributes like socio-demographic variables, environmental and conservation awareness, participation in village government committees, conservation activities, engagement in park-related tourism opportunities, protection of forests in buffer zone area, activities to reduce pressure on forest resources, assisting reserve managers to prevent illegal activities and forest patrol (Lai & Nepal, 2006; Nyaupane & Poudel, 2011; Regmi & Walter, 2016) have been further reported to support linkages between biodiversity conservation and local people.

Lai and Nepal (2006) also showed that though local people hold positive views of the measures necessary to achieve ecotourism goals, their intentions to engage in behaviours to support these measures do not entirely match with their positive views. These hindering factors include 'threats to livelihood by protected wildlife', 'restrictions of access to natural resources', 'poor relationships between protected area authorities' and 'lack of technical or financial support'. Nyaupane and Poudel (2011) too emphasized these factors as major causes of park/people conflict. All these findings thus highlight the importance of studying locals' perspectives, attitudes and behaviours towards ecotourism, either prior to its implementation or after its development. The present paper discusses local residents' post ecotourism development views in BWLS with a focus on homestays as an ecotourism product.

Homestays in ecotourism

Homestays may not represent a new form of tourism accommodation but they are arguably gaining increasing attention in academic literature as means to provide either direct/supplementary/alternative income to local communities, support local empowerment, alleviate poverty, attract (eco)tourists, showcase local cultural and natural heritage and interpersonally rich experience to visitors (Acharya & Halpenny, 2013; Coghlan, 2015; Gurung & Seeland, 2008; Knight & Cottrell, 2016; Kontogeorgopoulos, Churyen, & Duangsaeng, 2015; Regmi & Walter, 2016; Tran & Walter, 2014; Truong, Hall, & Garry, 2014). Lynch (2005, p. 528) defines homestays as follows:

> a specialist term referring to types of accommodation where tourists or guests pay to stay in private homes, where interaction takes place with a host and/or family usually living upon the premises, and with whom public space is, to a degree, shared.

Homestays appeal more often to international tourists searching for local lifestyle experience, novelty, personalized service and authentic/genuine social interactions with hosts (Kontogeorgopoulos et al., 2015; Mura, 2015; Wang, 2007), thus preferring local type of accommodation particularly in small, and often remote, rural communities. Singh (1991) suggested that by maintaining an indigenous touch, traditional settings and a vernacular architecture, host can offer an enjoyable experience for the guest. To be successful in providing a viable income to local communities and create rich and rewarding interpersonal and cultural experiences, the homestay programmes are often combined with other community-based tourism activities like camping, trekking, bird watching, the showcasing of traditional culture and festivals and so forth, either within the village or in adjoining areas. Like in Thrissur, Kerala, where homestay guests can enjoy cooking, yoga and *ayurveda* classes, plantation tours, hiking and treks, cultural tours, bird watching, wildlife safaris, cycling, access to local events and visits to local market (Paul, 2013). Homestays thus diversify the income opportunities for the villagers (Dutta, 2012) and generate interest in sustaining such nature-based activities. Regmi and Walter (2016) further emphasize on 'practice-based' learning process by locals while hosting homestay-based activities. Other studies show how linking traditional customs such as the sanctity of pilgrimage routes can also assist in preserving local cultural and natural heritage (e.g. Kaur, 1985; Singh & Kaur, 1983).

Precedents for the successful establishment of homestays are documented in the Jammu & Kashmir, Himachal Pradesh, Uttarakahnd, Sikkim and Kerala states of India. The development of these homestays have been linked to a range of local cultural activities; heritage and the natural capital component of a region; guiding and interpretation; forest patrols; development of promotional materials; equitable access for households across economic classes; crucial role of communities as tourism stakeholders; and towards conservation of natural resources and the environment (Anand, Chandan, & Singh, 2012; Dutta, 2012; Gangotia, 2013; Paul, 2013; Regmi & Walter, 2016; Sarkar & Sinha, 2015; Singh, Mal, & Kala, 2009). Regmi and Walter (2016) further highlight the connection between host's learning to cook a homestay meal to its local natural environment. Given these previous studies linking local support for ecotourism and conservation with perceived community benefits and the role of homestays as a direct community benefit from tourism, we might ask how homestay development affects the attitudes and behaviours of local communities towards ecotourism initiatives.

Despite the studies that link ecotourism, local communities, and community-benefit tourism initiatives to positive outcomes, the management of homestays themselves can be tricky and need careful management and planning. Kontogeorgopoulos et al. (2015) have questioned whether homestays are worth the risks for local communities with a desire to develop tourism initiatives. Several critics believe that the answer to this question is 'No', claiming that forcing hosts and guests to live together creates spatial, social and psychological impacts such as 'crowdedness, confusion, anxiety, ambiguity, privacy loss and degradation of quality of family life' (Oranratmanee, 2011, p. 46). These criticisms of homestay development notwithstanding, the present study assessed local residents' perspectives, attitudes and behaviours towards the ecotourism initiative within BWLS by focussing on the role of homestays as one specific form of CBE product.

Study setting

The development of homestays in BWLS

This study focuses on the homestays available in BWLS, located in villages and hub of lost culture and traditions as well as natural beauty (Paul, 2013). Homestays in Binsar are part of the Sanctuary's overall strategy to develop ecotourism opportunities within the area. Binsar is located in Kumaon Division of Uttarakhand, India at an altitude varying between 1500 and 2500 metres above sea level and covering an area of 47 km^2 (Figure 1). The area includes five villages (*Dalar*, *Risal*, *Satri*, *Gonap* and *Katdhara*) and, at the time of writing, was inhabited by 262 residents (comprising 57 households) (Table 1). In 1988, it was designated a wildlife sanctuary in order to safeguard the existing natural resources and prevent further forest degradation. The shift in Binsar's designation from a reserved forest (registered by the British Government in 1880) to a wildlife sanctuary did not occur without demur by the local communities living in and around this PA, in a similar manner to the 'voices of fear' during the declaration of the Gangotari National Park in Gharwal Division of Uttarakhand as mentioned by Chaturvedi (2002). Focusing primarily on protection and conservation of the ecosystem, the Regional Forest Department initially overruled local people's concern, prioritizing instead the increased risk of forest degradation and overuse of natural resources by villagers, for example, resin tapping, overgrazing, controlled forest fires, NTFP collection and illicit felling and fuel wood collection.

Following its designation as a sanctuary, Binsar formally started receiving tourists through its first entry gate built in 1999. The tourist activities permitted within BWLS include hiking on designated nature trails, wildlife spotting and bird watching, and enjoying Himalayan viewing points that attract an increasing number of nature tourists to Binsar every year. The five privately owned estates in Binsar (Table 2); one forest rest house (FRH) and one Kumaon Mandal Vikas Nigam (KMVN) tourist rest house were the only accommodations available within the sanctuary until the concept of homestays was introduced.

The designation of the wildlife sanctuary initially led to positive outcomes, as the biodiversity resumed its natural state. Gradually, negative impacts like instances of human–wildlife conflicts, forest fires and increased

Table 2. Distribution of villages and accommodations under different zones of Binsar PA.

Zone	Village name	Accommodation		
		Name	Type	Ownership
Core zone	None	None	–	–
Tourism zone		Kumaon Mandal Vikas Nigam (KMVN)	Hotel	Uttarakhand Government
		FRH	Guest house	Forest Department
		Mary Buden	Estate	Private
		Grand Oak Manor	Estate	Private
		Binsar Retreat	Resort	Private
		Nanda Devi	Estate	Private
		The Mountain Resort (Khali Estate)	Estate	Private
	Dalar	Eco-lodge	Lodge	Forest Department
		Mary Buden cottage	Cottage	Private
		Village-stay	Homestay 1	VWC
		Village-stay	Homestay 2	Private
		Village-stay	Homestay 3	Private
	Risal	Village-stay	Homestay 4	VWC
	Gonap	Village-stay	Homestay 5	VWC
		Village-stay	Homestay 6	Private
	Katdhara	Village-stay	Homestay 7	VWC
Buffer zone	Satri	Village-stay	Homestay 8	VWC

recreational impacts occurred. A total of 1763 domestic animals were killed by Leopards of BWLS during a 14-year period (Kala & Kothari, 2013). Furthermore, the loss of livelihood opportunities added to a growing fear of wild animals among villagers. Combined with a lack of basic amenities and little protection from wildlife, a mass exodus of villagers from the sanctuary was to follow. A new CBE initiative formed part of management strategies established to minimize the negatives impacts of the new sanctuary regulations while maximizing benefits to the local communities through ecotourism. The CBE initiative included the homestay programme along with other key ecotourism products such as nature guiding, trekking, enjoying Himalayan vistas, wildlife viewing, photography and bird-watching. The CBE initiative within the sanctuary aimed at, in part, to decrease villagers' migration, facilitate responsible job creation and create opportunities for better and more sustainable income sources.

Further development of CBE through the homestay programme occurred through a private–public partnership with the Village Ways Company (VWC). VWC represents a grassroots enterprise in Binsar, with an ongoing, close interaction with local community representatives and the Forest Department, who are key stakeholders in this context. The proposal to offer homestays within Binsar villages was one of the outcomes of this partnership and one homestay per village was built. VWC recommended building separate homestays rather than accommodating guests in existing homes in order to reduce the risk of jealousy, to protect villagers' privacy and to spread income among the community, thereby addressing many of the concerns raised by scholars such as Kontogeorgopoulos et al. (2015). Consequently, Village Tourism Committees (VTCs), locally known as Gram Paryatan Samitis (GPS), were formed.

The VTCs were established to build and manage (including determining rates and fees) these homestays. The VTCs represent all families agreeing to take part, and have a president, treasurer, secretary and vice-secretary, all unpaid and elected. Meetings are held regularly, and opportunities to work for the homestay in each community (cooking and cleaning) are rotated among the members. Nearly all families joined the VTCs, though some villagers were already employed in other sectors or received a pension from the state or the army. Homestays were built mostly by local artisans using local materials and traditional designs, on land leased by the VTCs from local landowners. The rent paid to the landowners is based on the number of guests staying each night. In addition, each village accrues an annual income of approximately one Lakh from the homestay activity and according to the VWC, 40% of the total income from homestays is spent on village administrative functioning. VWC helped villagers by covering construction costs through a mix of grants and interest-free loans. They also provided training in lodge management and guiding. The first homestay was opened to tourists in October 2006. Under the contract between each community and the landowner, the latter may buy the lodge from the VTC after 30 years. Based on the success of first five homestays built by the VWC, two villagers have started running private homestays. At present, eight homestays exist within the five villages of BWLS that function through community participation and according to the VWC tourist records, a total of 287 tourists, mostly from United Kingdom, stayed in Binsar's homestays in 2014–2015.

Methods

Following Yin's (2014) publication, the case-study method was selected for conducting the study at BWLS, which has undergone significant restructuring as part of its move towards ecotourism and has a strong focus on the inclusion of local communities within its ecotourism plan. Also, it explicitly aims to develop its CBE Initiative which seeks:

> to increase the effectiveness of conservation programmes in biologically important region by promoting private sector and community-based natural resource conservation, and by enabling communities to increase local tourism benefits, improve sustainability and compete more equitably with the regional tourism industry. (Wangchuk, 2002, p. 3)

Complete census for the study area was performed as our intent was to interview as many diverse villagers as possible. Data were collected using a mixed-methods approach, collecting background baseline information, interviews with members of the local villages and, field observations. Interview participants were approached with the assistance of the Forest Department, until all 57 heads of households in all five villages were included in the interviews.

Interview protocol and questionnaire

Based on Kvale (1996) method, face-to-face semi-structured interviews with a representative from every household in each village were conducted during January–February 2014. Interviews were read aloud in the Hindi language, manually transcribed, translated and coded by the lead author. Interviews with a single respondent lasted between 30 and 45 minutes and took place at the village-specific homestay itself (Table 2). Full participation was observed and the total number of completed interviews was 57. The interview included a range of questions on the respondents' village details, socio-demographic characteristics and finally, ecotourism-related questions. Respondents' village details assisted in describing the village's present status, the differences and similarities between and among these five villages (Table 3). As village details were kept non-categorical in nature, these were not subjected to statistical tests. The socio-demographic variables acted as explanatory variables for our study and are summarised in Table 4. The ecotourism-related questions that were asked during interviews include: type of engagement in ecotourism activities, planning and management, benefits from ecotourism (both personal and at a community level), barriers to participation (if any), perceptions and attitudes towards homestays and ecotourism and participating in ecotourism-directed behaviours. These questions regarding perceptions, attitudes and behaviour of Binsar villagers were formulated based on the preliminary field observations, the discussions with the villagers and the literature review process following Karanth and Nepal (2012) and Mintzer et al. (2015). Interview questions were presented as a mix of close-ended, yes/no questions, as well Likert-style questions on a three- and five-point scale to measure participation frequency and level of agreement

Table 3. Village details across five different villages within BWLS.

Village details	Village name					Mean/mode
	Dalar	Risal	Satri	Gonap	Katdhara	
Altitude (m)	1845	1711	1597	1908	1872	–
Total houses	14	16	11	11	35	17.4
Households interviewed (*n*)	09	09	03	10	26	–
Migrated	05	07	09	01	09	–
Access to electricity	Yes	No	No	No	No	No
Distance from metalled road						
Inside sanctuary road (km)	2	4	6	3	3	
Outside sanctuary road (km)	4	2	3	5	4	
Type of houses (Kutcha/Pucca)	Kutcha houses					
Type of energy consumption units (Fuelwood/LPG/both)	Fuelwood					

Table 4. Explanatory variables included in attitude assessment and percentage response of respondents across all households ($N = 57$) interviewed in five Binsar villages.

Socio-demographic variable	Village name					Mean/mode
	Dalar	Risal	Satri	Gonap	Katdhara	
Respondents gender (%)						Males
Male	100.0	89.0	100.0	90.0	77.0	
Female	0.0	11.0	0.0	10.0	23.0	
Household size (%)						4.5
01–05	11.1	66.7	100.0	70.0	88.5	
06–10	55.6	33.3	0.0	30.0	11.5	
>10	33.3	0.0	0.0	0.0	0.0	
Age (%)						52
18–30	11.1	0.0	0.0	10.0	0.0	
31–45	11.1	44.5	0.0	40.0	26.9	
46–64	44.5	33.3	66.7	40.0	57.7	
≥65	33.3	22.2	33.3	10.0	15.4	
Education (%)						Higher secondary
None	0.0	0.0	0.0	10.0	7.7	
Primary	0.0	11.1	0.0	20.0	7.7	
Lower secondary	66.7	22.2	100.0	40.0	23.2	
Higher secondary	11.1	44.5	0.0	30.0	53.8	
Undergraduate	22.2	11.1	0.0	0.0	3.8	
Post-graduate	0.0	11.1	0.0	0.0	3.8	
Occupation (%)						Subsistence agriculture
SA	44.5	44.5	33.3	80.0	53.8	
SA + daily wage labour	0.0	11.1	0.0	0.0	0.0	
SA + private sector job	33.3	22.2	33.3	20.0	11.5	
SA + public sector job	0.0	0.0	0.0	0.0	15.5	
SA + army retired (on pension)	22.2	22.2	33.3	0.0	19.2	
Land holding size (in Nali[a])						10–20 nali
01–10	55.6	44.5	0.0	50.0	38.5	
10–20	33.3	33.3	66.7	40.0	53.8	
>20	11.1	22.2	33.3	10.0	7.7	

[a]1 Acre = 20 nali (a land measurement unit practised in Uttarakhand).

Table 5. Respondents' attitude towards homestays in BWLS with interacting variables.

Attitude question	Responses ($N = 57$)			Interactions[a]	
				Explanatory variable	Participation behaviour
Oues 1 Do you like to participate in homestay programme?	Like 51(89.5%)	Neutral 6 (10.5%)	Dislike 0	Occupation (p-value = .039)	None
Oues 2 Do you think homestays provide important benefits?	Yes 42 (73.7%)	No 5 (8.8%)	Do not know 10 (17.5%)	None but seven reasons	Partnership (p-value = .012)
Oues 3 Do you think homestay contribute to ecotourism?	37 (64.9%)	6 (10.5%)	14 (24.6%)	None but five reasons	Partnership (p-value = .018) Cultural Programme (p-value = .049)
Ques 4 Will Binsar change if there are no homestays?	48 (84.2%)	4 (7.0%)	5 (8.8%)	None but five reasons	None

[a]Significant interactions with p-value < .050 were reported.

and finally, open-ended questions to allow respondents to elaborate on their answers. The interview questions also determined the willingness of villagers to contribute towards nature interpretation and secondly if they had faced human–wildlife interaction (HWI); and looked for significant correlations with their other attitudes, in line with the research of Zhang and Lei (2012). Furthermore, as the shift from natural resource-dependent income generation to tourism-related income opportunities occurred within BWLS, its villagers were asked to rate their opinion on preferred source of income, namely agriculture, daily wage and homestay using a Likert scale [1 = not important (minor), 2 = neither not important nor important and 3 = important (major)].

Attitude assessment

According to Rokeach (1968) 'attitude is defined as a relatively enduring organization of beliefs around an object or situation predisposing one to respond in some preferential manner' (p. 112). The villagers' attitude towards homestays was assessed through four close-ended questions (Table 5). Additionally, open-ended questions explored the reasons behind the answers provided for the attitude questions. The researcher avoided explaining ecotourism to the villagers to avoid any biasness in their response, thus obtaining a true representation of their perceptions and beliefs.

Behaviour assessment

The approach assessed if positive attitudes towards homestays were manifested as positive behaviours towards the sanctuary and its management objectives. For this, 14 specific ecotourism-directed behaviours of villagers' were quantified (Table 8). A three-point Likert scale – never, sometimes, and frequently – was used to quantify participation frequency towards ecotourism supporting activities. We expected that if positive attitudes towards homestays affect participation in other ecotourism supporting activities, villagers who expressed positive attitudes would be more willing to contribute to ecotourism objectives. Through open-ended questions, the interviews also investigated the reasons for their participation. To examine the potential influence of the homestay programme on the behaviour (participation), we also asked the villagers 'What would you do if there was no homestay programme?' and investigated for attitudes correlating to it.

Data analysis

We analysed the closed-ended responses using standard non-parametric statistical tests in SPSS. For the open-ended questions, a thematic content analysis was done. Fisher's Exact Test was used to determine which explanatory variables were significantly correlated with positive attitudes (at 5% alpha level of significance). The Fisher's Exact Test was appropriate given the categorical nature of the explanatory variables and the small sample sizes (as expected frequency was less than five). Some explanatory variables were reclassified into a maximum of three categories in order to facilitate analysis in SPSS. Regarding behaviour assessment, the positive responses were thus reported as sum of 'sometimes' and 'frequently' percentage outcomes. Finally, data were analysed to determine significant correlation between any two variables, if any, and to make interpretations based on obtained probability value 'p-value'.

Results

Homestays of Binsar

Homestays represent 47% ($n = 8$) of all tourism accommodations available within BWLS (Figure 2), of which three (37.5%) are privately owned and five (62.5%) are owned by VWC. The five VWC homestays function with coordination of 57 families among whom the benefits are shared equally. The three private homestays, two in

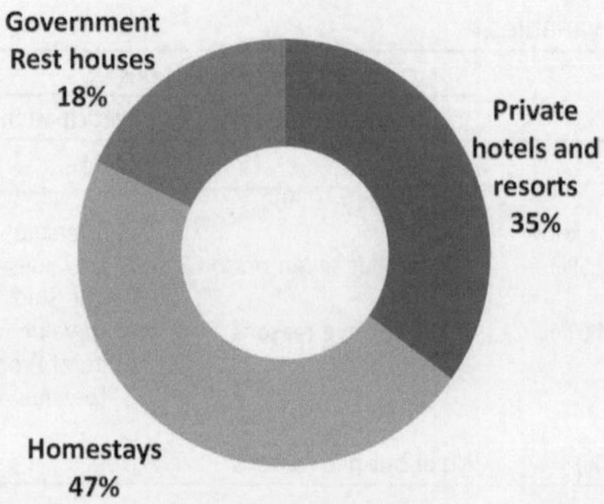

Figure 2. Percentage distribution of different types of tourism accommodations within BWLS.

Dalar village and one in *Gonap* village (Table 2), are owned independently by two families. Also, the privately owned homestay in *Gonap* village (Figure 3) represents the only homestay where the guests get the opportunity to reside in the owner's house, albeit in separately constructed section of the home. The two privately owned homestays in Dalar village consist of one bedroom (for two persons) and two bedrooms (for four persons), respectively, as well as a kitchen and a bathroom. Each VWC homestay is designed in similar fashion comprising of three bedrooms (six beds), one dining hall and a kitchen (Figure 4). No more than six guests per village visit is allowed according to the VWC norms. According to the Binsar tourist arrival records, only 4.6% of the total overnight tourists opted for homestay option in 2014–2015.

The practice of homestays provision in BWLS includes the participation of all villagers (Figure 5), who contribute directly and indirectly to ecotourism initiatives by providing land for construction of homestay as well as use of traditional knowledge and techniques of construction and manual labour. Villagers provide locally grown vegetables, dairy products from domesticated livestock as well as prepare meals for visitors. They will also offer cultural performances, reciting folklore and showcasing local practices in response to tourists' requests. Finally, all maintenance and security are undertaken by villagers. Figure 5 reflects ecotourism functioning within BWLS, highlighting community organisation, participating activities and its types. The following part of the results section presents data from the interviews on ecotourism practice occurring in BWLS, with a specific emphasis on community participation activities.

Figure 3. (a) Location of *Gonap* village within BWLS, (b) private homestay within *Gonap* village and (c) tourists on a guided nature trail within sanctuary.

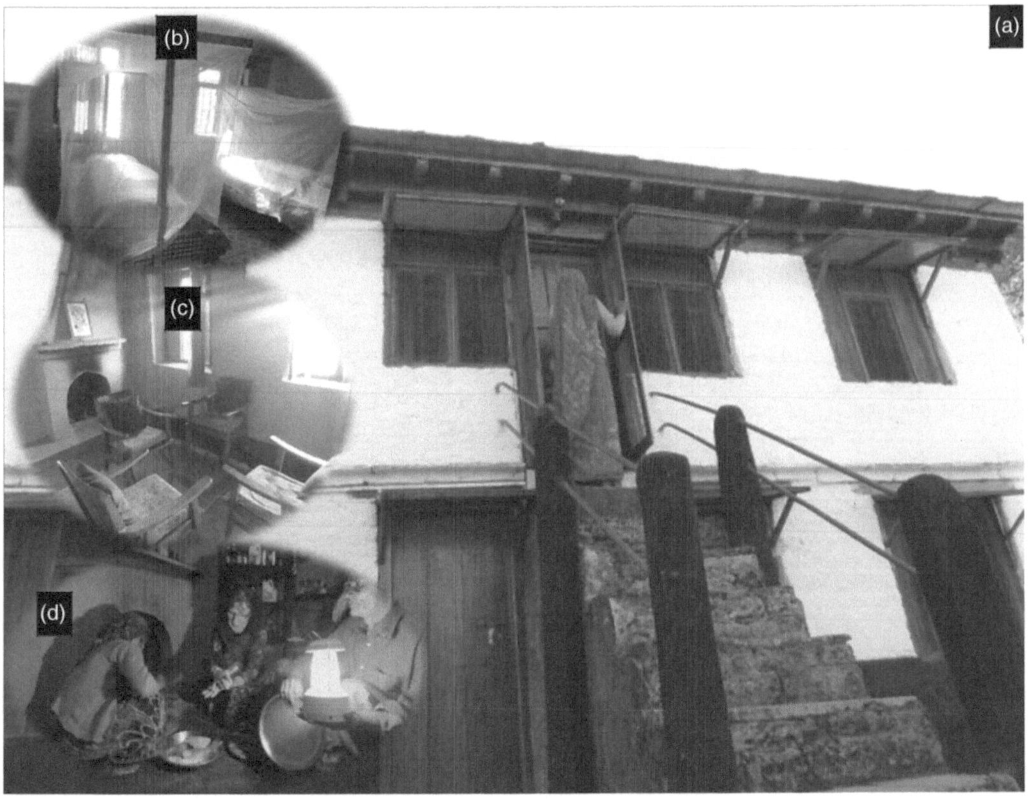

Figure 4. (a) VWC homestay within *Dalar* village, (b) a bedroom within a homestay, (c) a living room within a homestay and (d) tourist engaged in cooking activity.

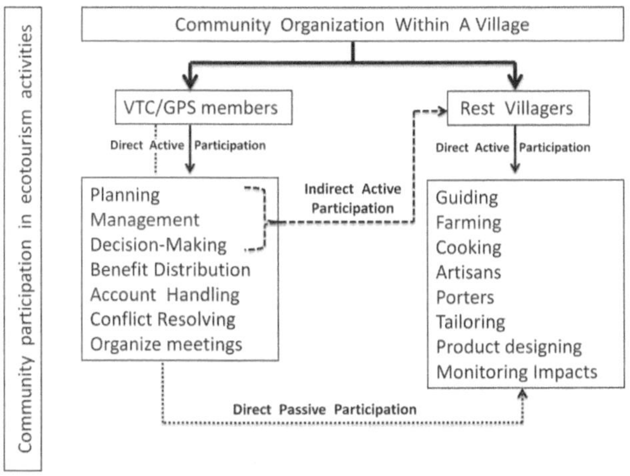

Figure 5. Internal community structure model for ecotourism functioning in BWLS.

Overview of participants

The average age of the participants was 52 (SD = 12.21) ranging from 29 to 75 (Table 4). Forty-nine of the total interviewees were male. We targeted household heads because the preliminary interviews suggested that they were actively involved in planning and management activities, were mainly involved with the homestay development programme, had experienced changes if any and were easily accessible in their respective homes (others were either below 18 years or working). Though women were engaged in their daytime household activities like farming, cattle grazing, milking cow and cooking, most were reluctant to be interviewed. However, eight women showed their interest in being interviewed on familiarization with the lead author.

Villagers' perceptions and attitudes

The majority of the respondents ($n = 41$) perceived ecotourism as an activity where the 'visitors could enjoy nature' (Table 7). While 27 respondents considered ecotourism as an 'income-generating opportunity' and only 4 respondents mentioned the term 'eco-friendly' while describing it. In describing the current sources of income (Table 6) within the sanctuary, villagers indicated subsistence agriculture (SA), daily wage labour and homestays with its associated activities as the only available opportunities. Out of these, the weighted mean readings indicated homestays as their preferred source of income (Figure 6). Thirty (52.6%) respondents reported homestays as the majorly preferred source of income, 34 respondents (59.6%) reported daily wage labour as the neutral source of income and 39 respondents (68.4%) reported agriculture as the minor source of income

Table 6. Interactions among different variables observed from respondents interview in BWLS.

Question	Response (N = 57)			Interacting variable	Statistic: Fishers Exact value	Statistic: p-Value
Whether faced HWI	Yes 48 (84.2%)	No 09 (15.8%)	–	(1) Agriculture as minor source of income	25.736	.000
				(2) Land holding size	6.537	.036
Willing to contribute to nature interpretation	Willing 41 (71.9%)	Not willing 16 (28.1%)	–	(1) Household size	–	.010
				(2) Attitude Ques 1[a]	–	.046
				(3) Attitude Ques 3[a]	5.130	.050
Preferred source of income:	Minor	Neutral	Major			
- Agriculture	39 (68.4%)	15 (26.3%)	03 (5.3%)	(1) HWI	–	–
- Daily wage	12 (21.1%)	34 (59.6%)	11 (19.3%)	(1) Age	8.575	.042
- Homestay	06 (10.5%)	21 (36.8%)	30 (52.6%)	None	–	–

*Attitude questions are given in Table 5.

Table 7. Perceptions and reasons reported by more than one respondent to open-ended questions.

Open-ended question	Perceptions/reasons	Percent of respondents
How do you perceive ecotourism?		(*n* = 55)
	It is for visitors to enjoy nature	41
	It generates job opportunities	27
	It helps to protect nature	12
	It is for wildlife protection	7
	It is eco-friendly tourism	4
Why do you consider homestays to be important		(*n* = 42)
	It is our livelihood	39
	It prevents migration	25
	It assists in cultural exchange	17
	It increases community-brotherhood	6
	It is cheaper than other hotels	5
	It has improved our standard of living	4
	Familiarization with technological advancements	2
Why do you believe homestays contribute to ecotourism		(*n* = 37)
	Generate responsibility to protect natural resources for which guests pay	33
	People visit us to experience nature and silence in homely atmosphere	25
	Give sense of ownership	20
	More engagement in homestay related activities refrain from illegal extraction of forest produce	12
	Created guiding jobs	8
What would be the possible outcomes if there was no homestay programme?		(*n* = 52)
	Migration	40
	Increase in human–wildlife conflicts	39
	Insecure old age	25
	Lack of basic facilities	18
	Lesser solid waste generation	3

within BWLS (Table 6). Furthermore, daily wage labour as neutral source of income was found to be based upon 'age' variable (Fisher's Exact Test, *p*-value = .042) and was maximally (*n* = 18) reported by age group of 30–50 years. Thus, these findings highlighted homestays as the BWLS villagers' preferred source of income (Table 6).

In addition, 48 (84.2%) respondents reported to have faced HWIs (Table 6) either in the form of livestock depredations or as crop-raiding over the years and nearly all the villagers highlighted this threat as a hindrance to their agricultural practices. Furthermore, it was explored that 81.6 % (*n* = 39) of these 48 respondents had actually reported agriculture as their minor source of income, resulting in a statistically significant correlation between income type and perceptions of human–wildlife conflicts (Fisher's Exact Test, *p*-value = .000). Furthermore, the difference regarding whether a villager had faced human–wildlife conflict or not was also based upon the difference in their land holding size (Fisher's Exact Test, *p*-value = .036); and it was explored that 66.7% (*n* = 32) of these 48 respondents having small landholdings had face such conflicts.

Next, 51 respondents replied that they liked participating in the homestay programme (89.5%), 6 (10.5%) gave a neutral response and none disliked being in the programme (Table 5). The only explanatory variable that was significantly correlated with opinion was 'Occupation' (Fisher's Exact Test, *p*-value = .039). On further exploration, it was found that out of these 51 respondents, villagers who were only engaged in

Table 8. Self-reported participation frequencies towards ecotourism supporting and opposing activities.

Participation in	Percent response[a] (N = 57)
Wildlife census and ecotourism impact monitoring	26.3
Reporting of natural incidences and anthropogenic activities	54.4
NTFP collection	8.8
Hunting for wildlife	0.0
Burning of solid waste	24.6
Homestay designing	87.7
Management planning and decision-making process	94.8
Partnerships	91.2
Cultural programmes for visitors	91.2
Infrastructural development	56.2
Designing visual aids for destination promotion	28.0
Designing of ecotourism products	94.8
Opportunities in promotional videos and advertisement	93.0
Trainings and workshops	100.0

[a]Figures indicate % of respondents reporting 'frequent' and 'sometimes' participation in listed activities.

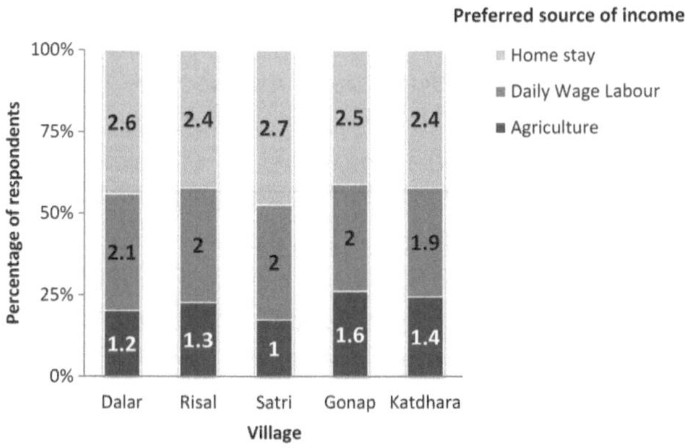

Figure 6. Villagers' response to preferred source of income within BWLS.

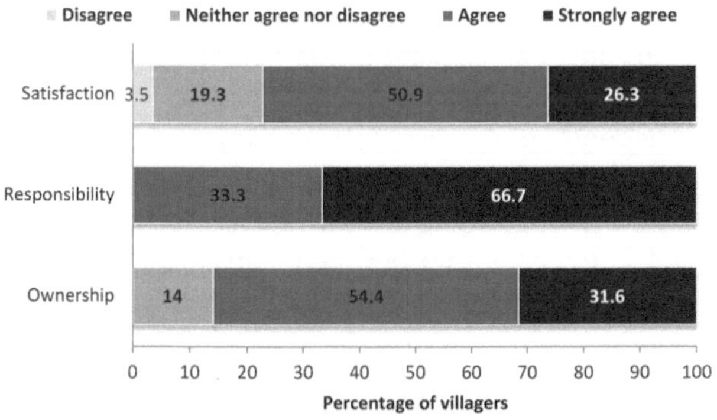

Figure 7. Villagers' level of agreement towards satisfaction, responsibility and ownership attitudes.

subsistence farming ($n = 31$; 60.8%) liked to participate in the homestay programme more than villagers engaged in other jobs besides subsistence farming. No other extraneous variable was observed to have an effect on the villager' positive attitude towards the homestay programme.

Considering the benefits of participating in the homestay programme, we asked respondents whether homestays provide important benefits for life in Binsar. Forty-two respondents (73.7%) replied affirmatively (Table 5) and provided seven different reasons (Table 7) for their answer, focusing on livelihoods ($n = 39$), prevention of migration ($n = 25$) and improved lifestyle ($n = 4$). Additionally, the villagers were asked if the homestay programme contributes towards ecotourism; 64.9 % ($n = 37$) of respondents responded favourably (Table 5) and provided five reasons (Table 7) for their response, focusing on generation of responsibility to protect natural resources ($n = 33$), sense of ownership ($n = 20$) and refrain from illegal extraction of forest produce ($n = 12$). None of the explanatory variables was observed to have an effect on the positive attitude of villagers towards the homestay programme.

Furthermore, respondents were asked if they believe that Binsar would change if the homestay programme did not exist. Forty-eight respondents replied affirmatively (84.2%), while four (7%) did not believe it would change (Table 5). The majority of respondents suggested that the absence of the homestay programme would lead to a negative change because there would be no other alternative income opportunity left for them, and they would be forced to migrate (Table 7) out of the Sanctuary ($n = 40$). Some respondents ($n = 3$) noted that it would create lesser solid waste generation and thus reduce the challenge of managing it (Table 7). Whereas others reported 'increasing human–wildlife conflicts' ($n = 39$), 'old age security' ($n = 25$) and 'lack of basic facilities' ($n = 18$) as possible outcomes of the homestay programme absence within BWLS. We did not find any statistically significant correlation between this attitude variable and explanatory variables (Table 7).

Homestays and local guardianship

We measured villagers' ($N = 57$) level of agreement on the various outcomes of the homestay programme using a five-point Likert scale (strongly disagree 1 to strongly agree, 5). Figure 7 reports that 50.9% ($n = 29$) of the respondents agreed that they are 'satisfied working with the homestay programme', 66.7% ($n = 38$) of the villagers strongly agreed with the statement 'I feel more responsible towards the natural resources now' and 54.4% ($n = 31$) agreed with the statement that the 'Homestay programme has generated ownership feeling'. Majority of responses suggest a positive response of villagers towards the homestay programme and arguably indicate an interest in being guardians of their natural environment.

Willingness to contribute towards nature interpretation

Out of 57 respondents, 41 respondents (71.9%) reported 'willing' response and this willingness was found to be based upon three interacting variables (Table 6). First, the household size (Fisher's Exact Test, p-value = .010), within which 82.5% ($n = 33$) of the total willing respondents belonged to a household of up to five individuals. Second, towards having either positive or neutral attitude towards participation in the homestay programme (Fisher's Exact Test, p-value = .046), within which 76.5% ($n = 39$) showed positive attitude towards participation. And third, towards the attitude whether the homestay programme contributes to ecotourism or not (Fisher's Exact Test, p-value = .050), within which 73% ($n = 27$) reported that homestay contributes to ecotourism. When further asked how they might contribute, those 'willing' respondents suggested 'designing' and 'use of signage boards' to enhance the interpretation component of ecotourism within BWLS.

Behaviour through participation in ecotourism-supporting activities

The attitude, whether a respondent considered the homestay programme important or not was found to be significantly correlated (Table 5) with whether or not villagers were engaged in any 'partnership' behaviour (Fisher's Exact Test, p-value = .012). Unexpectedly, 10.7% ($n = 3$) of the villagers reporting homestays as 'not important' and 32.1% ($n = 9$) as 'don't know' were found to be frequently engaged in various partnerships. On further discussions, it was found that these partnerships exist at small-scale regional levels, for example, with local travel tour operators and transport providers; and selling locally produced honey and *Rhododendron* flower juice either directly to visitors on request or indirectly through souvenir shops.

Behaviours favouring social aspects of ecotourism like 'engagement in partnerships' (Fisher's Exact Test, p-value = .018) and 'involvement in cultural programmes for visitors' (Fisher's Exact Test, p-value = .049) were found to be significantly correlated with their response to whether the homestay programme contributes to ecotourism or not (Table 5). However, unexpectedly, behaviours (Table 8) favouring conservation aspect of ecotourism, such as engagement in 'wildlife census and ecotourism impact monitoring', 'reporting of natural incidences and anthropogenic activities' and participation in 'trainings and workshops' besides non-engagement in 'NTFP collection', 'hunting for wildlife' and 'burning of solid waste', were not found to be significantly correlated with this attitude variable. Besides this, none of the participation behaviour variables was further found to be significantly correlated with the reported respondents' attitude towards 'homestay absence'.

Discussion

Considering homestays as a CBE product, this case study assessed whether homestays inclusion in an ecotourism strategy stimulates local guardianship of natural resources and encourages support for ecotourism objectives. Through assessment of attitudes and behaviour of local villagers towards homestays and their contribution to ecotourism, the findings challenge the notion that CBE is inevitably fraught with tension within community structures and between concerned stakeholders, such as natural resource managers and tourism businesses. In contrast to the fear that a national park declaration will hinder local participation as reported by Chaturvedi (2002), the idea of the homestay programme within BWLS was appealing to nearly all its inhabitants, who actively participated in the programme. While anecdotal evidence suggests that the link between local communities and ecotourism has been a success for BWLS, this study has explored how various conservation issues (the dependency of local people on forest resources; anthropogenic pressures [tourism]; lack of conservation awareness and values of Kumaon Himalayas among visitors and/or local people) noted in the years after the sanctuary was established (Ilyas, 1998) have evolved with the establishment of the CBE-homestay over the course of one generation of villagers (a 27-year timeline).

Our findings suggest that Binsar villagers were positive towards participating in the homestay programme due to the direct income opportunities, as indicated by respondents' preference of homestays as their major source of income. This supports Nyaupane and Poudels' (2011) view on 'direct-linkage scenario' developing interrelationship between conservation and surrounding communities through mutually beneficial relationship. This attitude in turn was found to be based upon the 'human–wildlife interaction' and 'landholding size' variables. These variables may be categorized under 'intervening variables' deciding participation of villagers as discussed by Lai and Nepal (2006) and could be the subject of further exploration and research. Behaviours like 'engagement in partnerships' and 'cultural programmes for visitors' were the only participating variables found to be significantly correlated with the attitudes of respondents agreeing that 'homestays provide important benefits' and 'homestay contribute to ecotourism'. Furthermore, the 'partnerships' behaviour exhibited by BWLS residents (91.2%) reflects their intentions to create linkages between rural tourism set-up with other relevant sectors, in particular through travel and tour operators, and local produce (like honey and *Rhododendron* flower juice) supply. This may emphasize the leadership role the locals are practising in BWLS to extend tourism business and promote followership amongst local rural tourism small-scale business (Haven-Tang & Jones, 2012; Truong et al., 2014) and thus gain popularity. It may also contribute to locals' self-esteem, self-confidence and self-determination and, control of ecotourism development and management (Nyaupane & Poudel, 2011; Regmi & Walter, 2016).

The majority of the villagers understood ecotourism as a nature-based activity and strongly believed that with engagement in the homestay programme, many felt that they had gained a sense of ownership and felt responsible towards maintaining Binsar's natural setting. All these interactions arguably indicate villagers' interest in being guardians of their natural environment. Furthermore, the purpose shown by Binsars' respondents towards protecting the local environment gives an insight to villagers' attitude towards conservation, as argued by Zhang and Lei (2012). Our findings also support Lai and Nepal's (2006) review of variables that may predict communities' attitudes or behaviours towards conservation of PAs. Community's attitude can be either favourable or unfavourable towards conservation based upon these variables. Favourable as despite majority of Binsar villagers have faced human–wildlife conflict, yet they support the homestay programme, show willingness to contribute towards nature interpretations, have reduced NTFP collection and, none practises hunting. Unfavourable as despite Binsar villagers' belief that homestays contribute to ecotourism, only few actually participated in wildlife census and ecotourism impact monitoring and, still practise burning of solid waste within the sanctuary. Thus, in support to Lai and Nepal's (2006) view, though community shows an overall positive attitude towards ecotourism, they need not always engage in behaviours to support it. Local support for these tasks is invaluable, as villagers can help in locating faunal species using their local knowledge and therefore assist forest officers and wildlife biologists in species counts. Besides this, they can also assist reserve managers in forest patrol and in ecological monitoring and, to prevent illegal activities, so as to achieve the goals of ecotourism to protect the natural environment as suggested by Lai and Nepal (2006), Nyaupane and Poudel (2011), Regmi and Walter (2016) and Zhang and Lei (2012). Though other positive behaviours like reporting of natural incidences and anthropogenic activities, attending trainings and workshops, and participation in management planning and decision-making processes related to CBE are practised by many villagers, it was not found to be significantly correlated with any of the positive attitude variables.

The benefits reported from participating in the homestay programme from our study like 'prevention of migration' ($n = 25$) and 'improvement in living standards' ($n = 4$) were found in Binsar. These were observed as potential outcomes of participation in ecotourism as reported by Kala (2013) and Chaturvedi (2002) in similar studies. Simultaneously, owing to extremely low population density within BWLS, it may have become likely to put enough money into local communities to affect a change in local attitudes towards conservation, as proposed by Bookbinder et al. (1998). Binsar villagers have shown their keen interest in such CBE ventures by participating in ecotourism-directed activities and contribute to ecotourism objectives through an adaptive process. Similar studies by Karanth and Nepal (2012) in Nepal and India also revealed that most local residents involved in community-based tourism initiatives had favourable attitudes towards PAs. This in turn provides a strong incentive to educate and encourage local resident participation in conservation and protection efforts. Bhattacharya, Banerjee, and Saksena (2003) also reported that 75% of the villagers surrounding Kerwa and Van Vihar National Park were willing to be associated with ecotourism and its related activities, but the study did not indicate reasons behind such willingness. On the other hand, Stem, Lassoie, Lee, Deshler, and Schelhas (2003) showed that benefits derived from ecotourism may not change local people's attitudes towards conservation.

Our study highlighted that some further opportunities may exist to consolidate local residents' support for the sanctuary by capitalising on latent opportunities. BWLS villagers are aware that homestays would survive only if the natural habitat surrounding homestays is protected. Also, they are willing to protect the destination, but face difficulty in showing their actual participation towards achieving ecotourism objectives. This is an area that can be addressed by new management practices initiated by concerned stakeholders. Capitalising on the abundant traditional aspects represents another latent opportunity in Binsar. Thus, the situation supports the opinion with Karanth and Nepal (2012) that in order to sustain tourism in PAs, it would require sharing of benefits with local people and building support among private enterprises for conservation initiatives.

In order to maximize conservation benefits, forest officials, local government and tourism experts shall have to play an important role to manage the relationship between the sanctuary ecosystem and residents so that the latter can be fully benefited from ecotourism-directed homestays. In practical terms, with the declaration of the wildlife sanctuary, residents have experienced more frequent negative HWIs. This has led to a perception that even subsistence farming is becoming unfavourable. The findings thus suggest that in order to support and benefit from ecotourism initiatives within the sanctuary, these interactions must be managed and local communities protected in this context. Secondly, generation of solid waste and its management pose another threat to the locals, as they themselves find unable to handle this waste on their own. Current waste management practices not only affect the surrounding environment but also may impact the wildlife directly or indirectly. Leadership efforts of tourism experts and forestry officers to provide training for local residents on solid waste management should therefore be encouraged and supported. Kala (2013) suggested the use of holdings on the road and vantage points depicting the ill effects of littering can help in spreading awareness among the tourists and locals to use dustbins. This is supported by Binsar villagers' interest on using and designing of signage boards, presently existent as willingness only. Encouraging participation for this activity would support the interpretation component of ecotourism, while developing the untapped creative skills of local communities. In a similar manner, designing of an ecotourism monitoring and management plan and, conservation supporting activities are important in Binsar, so that the villagers are as attracted to it as they are to homestays. While the results of study point towards a largely positive relationship between local residents, the sanctuary, the forest department and the VWC, anticipating and responding through responsible leadership will sustain this positive relationship as ecotourism and homestays continue to develop in BWLS.

Conclusion

Within the context of BWLS, homestays appear to contribute directly and indirectly to the local community, and thereby encourage support for ecotourism objectives within the sanctuary. Thus, this case study supports the notion that homestays can help the viability of CBE because it has the capability to generate sustained income for local individuals who struggle with remote area hardship and where other job opportunities are lacking. Homestays therefore present the potential to generate a sense of ownership and responsibility towards the natural setting and contribute to conserving the local cultural and natural heritage. Finally, once the locals receive tangible benefits and are involved in conservation policies, they consider themselves as stakeholders in the process and support the conservational policies. In this sense, we can develop a better understanding of how the inclusion and management of homestays as part of an ecotourism initiative within a PA might provide a successful contribution to community development through tourism.

Acknowledgements

The authors are grateful to the three anonymous reviewers for their insightful comments. The researchers duly acknowledge the valuable inputs provided by the Village Ways Company, the logistic support provided by the Regional Forest Department of Uttarakhand, India, and the GIS guidance provided at Indira Gandhi Conservation and Monitoring Centre (IGCMC) of WWF-India.

Disclosure statement

No potential conflict of interest was reported by the authors.

Funding information

The researcher duly acknowledges the financial assistance provided by GGS Indraprastha University, New Delhi, India, under the Indraprastha Research Fellowship (IPRF), Award letter number (GGSIPU/IPRF/2012/85) and the Dean USEM.

References

Acharya, B. P., & Halpenny, E. A. (2013). Homestays as an alternative tourism product for sustainable community development: A case study of women-managed tourism product in rural Nepal. *Tourism Planning and Development, 10*(4), 367–387.

Anand, A., Chandan, P., & Singh, R. B. (2012). Homestays at Korzok: Supplementing rural livelihoods and supporting green tourism in the Indian Himalayas. *Mountain Research and Development, 32*(2), 126–136.

Bhattacharya, A. K., Banerjee, S., & Saksena, V. (2003). Local initiatives to localize ecotourism: An exploratory study in Kerwa, Van Vihar National Park catchments. *Tourism Recreation Research, 28*(1), 97–102.

Birdlife International. (2016). Important bird and biodiversity area factsheet: Binsar Wildlife Sanctuary. Retrieved from http://www.birdlife.org on 07/02/2016

Bookbinder, M. P., Dinerstein, E., Rijal, A., Cauley, H., & Rajouria, A. (1998). Ecotourism's support of biodiversity conservation. *Conservation Biology, 12*(6), 1399–1404.

Brandon, K. (2001). Moving beyond integrated conservation and development projects (ICDP's) to achieve biodiversity conservation. In D. R. Lee & C. B. Barrett (Eds.), *Tradeoffs or synergies? Agricultural intensification, economic development and the environment* (pp. 417–432). New York, NY: CABI Publishing.

Brandon, K., & Margoluis, R. (1996). Structuring ecotourism success: Framework for analysis. In E. Malek-Zadeh (Ed.), *The ecotourism equation: Measuring the impacts* (pp. 28–38). New Haven, CT: Yale University School of Forestry and Environmental Studies.

Chaturvedi, G. (2002). Ecotourism in Gangotri region of the Garhwal Himalayas. *Tourism Recreation Research, 27*(3), 41–51.

Cobbinah, P. B. (2015). Local attitudes towards natural resources management in rural Ghana. *Management of Environmental Quality: An International Journal, 26*(3), 423–436.

Coghlan, A. (2015). Prosocial behaviour in volunteer tourism. *Annals of Tourism Research, 55*, 46–60.

Das, M., & Chatterjee, B. (2015). Ecotourism: A panacea or a predicament? *Tourism Management Perspectives, 14*, 3–16.

Drumm, A., & Moore, A. (2005). *Ecotourism development – a manual for conservation planners and managers: An introduction to ecotourism planning* (Vol. 1, 2nd ed.). Arlington, VA: The Nature Conservancy.

Dutta, P. K. (2012). *Guidelines for promotion of Homestays in Arunachal Pradesh*. Tezpur: Western Arunachal Landscape Programme, WWF-India.

Fennell, D. A. (2015). *Ecotourism* (4th ed.). New York, NY: Routledge.

Gangotia, A. (2013). Home stay scheme in Himachal Pradesh: A successful story of community based tourism initiatives (CBTIS). *Global Research Analysis, 2*(2), 206–207.

Gurung, C. P. (1995). People and their participation: New approaches to resolving conflicts and promoting cooperation. In J. M. McNeely (Ed.), *Expanding partnerships in conservation* (pp. 223–233). Washington, DC: Island Press.

Gurung, D. B., & Seeland, K. (2008). Ecotourism in Bhutan – extending its benefits to rural communities. *Annals of Tourism Research, 35*(2), 489–508.

Haven-Tang, C., & Jones, E. (2012). Local leadership for rural tourism development: A case study of Adventa, Monmouthshire, UK. *Tourism Management Perspectives, 4*, 28–35.

Heinen, J. T. (1996). Human behavior, incentives and protected area management. *Conservation Biology, 10*, 681–684.

Ilyas, O. (1998). *People and protected areas – the case of Binsar Wildlife Sanctuary*. New Delhi: WWF-India.

Kala, C. P. (2013). Ecotourism and sustainable development of mountain communities: A study of Dhanolti Ecopark in Uttarakhand state of India. *Applied Ecology and Environmental Sciences, 1*(5), 98–103.

Kala, C. P., & Kothari, K. K. (2013). Livestock predation by common leopard in Binsar Wildlife Sanctuary, India: Human–wildlife conflicts and conservation issues. *Human–Wildlife Interactions, 7*(2), 325–333.

Karanth, K. K., & Nepal, S. K. (2012). Local residents' perception of benefits and losses from protected areas in India and Nepal. *Environmental Management, 49*, 372–386.

Kaur, J. (1985). *Himalayan pilgrimages and the new tourism*. New Delhi: Himalayan Books.

Knight, D. W., & Cottrell, S. P. (2016). Evaluating tourism-linked empowerment in Cuzco, Peru. *Annals of Tourism Research, 56*, 32–47.

Kontogeorgopoulos, N., Churyen, A., & Duangsaeng, V. (2015). Homestay tourism and the commercialization of the rural home in Thailand. *Asia Pacific Journal of Tourism Research, 20*(1), 29–50.

Kvale, S. (1996). *Interviews – an introduction to qualitative research interviewing*. Thousand Oaks, CA: Sage.

Lai, P.-H., & Nepal, S. K. (2006). Local perspectives of ecotourism development in Tawushan Nature Reserve, Taiwan. *Tourism Management, 27*, 1117–1129.

Lindberg, K., & Hawkins, D. E. (1999). *Ecotourism – a guide for planners and managers* (Vol. 1). Dehradun: Natraj.

Lynch, P. (2005). The commercial home enterprise and host: A United Kingdom perspective. *International Journal of Hospitality, 24*(4), 533–553.

Mehta, J., & Heinen, J. (2001). Does community-based conservation shape favorable attitudes among locals? An empirical study from Nepal. *Environment Management, 28*, 165–177.

Mintzer, V. J., Schmink, M., Lorenzen, K., Frazer, T. K., Martin, A. R., & da Silva, V. M. F. (2015). Attitudes and behaviors toward Amazon River dolphins (*Inia geoffrensis*) in a sustainable use protected area. *Biodiversity Conservation, 24*, 247–269.

Mura, P. (2015). Perceptions of authenticity in a Malaysian homestay – A narrative analysis. *Tourism Management, 51*, 225–233.

Nyaupane, G. P., & Poudel, S. (2011). Linkages among biodiversity, livelihood, and tourism. *Annals of Tourism Research, 38* (4), 1344–1366.

Okazaki, E. (2008). A community-based tourism model: Its conception and use. *Journal of Sustainable Tourism, 16*(5), 511–529.

Oranratmanee, R. (2011). Re-utilizing space: Accommodating tourists in homestay houses in northern Thailand. *Journal of Architectural/Planning Research and Studies, 8*(1), 35–54.

Paul, H. K. (2013). Home stays for the development of tourism in Thrissur district. Research paper in proceedings of UGC sponsored national seminar on "Service sector dynamism and economic growth in India: Prospects and problems" held at Post Graduate Department of Commerce and Management Studies, Sri C Achutha Menon Government College, Thrissur, Kerala from 18th–19th December 2013, pp. 100–109.

Rastogi, A., Badola, R., Hussain, S. A., & Hickey, G. M. (2010). Assessing the utility of stakeholder analysis to protected areas management: The case of Corbett National Park, India. *Biological Conservation, 143*, 2956–2964.

Regmi, K. D., & Walter, P. G. (2016). Conceptualising host learning in community-based ecotourism homestays. *Journal of Ecotourism*. doi:10.1080/14724049.2015.1118108

Reimer, J. K., & Walter, P. (2013). How do you know it when you see it? Community-based ecotourism in the Cardamom Mountains of southwestern Cambodia. *Tourism Management, 34*, 122–132.

Rokeach, M. (1968). *Beliefs, attitudes, and values: A theory of organization and change*. San Francisco, CA: Jossey-Bass.

Sarkar, R., & Sinha, A. (2015). The village as a social entrepreneur: Balancing conservation and livelihoods. *Tourism Management Perspectives, 16*, 100–106.

Singh, R. B., Mal, S., & Kala, C. P. (2009). Community responses to mountain tourism: A case in Bhyundar Valley, Indian Himalaya. *Journal of Mountain Science, 6*, 394–404.

Singh, T. V. (1991). The development of tourism in the mountain environment, the problem of sustainability. *Tourism Recreation Research, 16*(2), 1–11.

Singh, T. V., & Kaur, J. (1983). *Himalayas, mountain and men.* Lucknow: Print House.

Sinha, B. C., Qureshi, Q., Uniyal, V. K., & Sen, S. (2012). Economics of wildlife tourism – contribution to livelihoods of communities around Kanha Tiger Reserve, India. *Journal of Ecotourism, 11*(3), 207–218.

Stem, C., Lassoie, J., Lee, D., Deshler, D., & Schelhas, J. (2003). Community participation in ecotourism benefits: The link to conservation practices and perspectives. *Society and Natural Resources, 16*, 387–413.

Stone, M. T. (2015). Community-based ecotourism: A collaborative partnerships perspective. *Journal of Ecotourism*. doi:10.1080/14724049.2015.1023309

Tran, L., & Walter, P. (2014). Ecotourism, gender and development in northern Vietnam. *Annals of Tourism Research, 44*, 116–130.

Truong, V. D., Hall, C. M., & Garry, T. (2014). Tourism and poverty alleviation: Perceptions and experiences of poor people in Sapa, Vietnam. *Journal of Sustainable Tourism, 22*(7), 1071–1089.

Wang, Y. (2007). Customized authenticity begins at home. *Annals of Tourism Research, 34*(3), 789–804.

Wangchuk, R. (2002). Report on "a learning tour of the CBN (Corbett, Nainital and Binsar) eco-tourism initiative sites by villagers from Hemis National Park and the surrounding area". Field series document no. 5. Ladakh: Snow Leopard Conservancy Leh.

Wells, M. P., & Brandon, K. E. (1993). The principles and practice of buffer zones and local participation in biodiversity conservation. *Ambio, 22*(2–3), 157–162.

Yin, R. K. (2014). *Case study research: Design and methods* (5th ed.). Beverly Hills, CA: Sage.

Zhang, H., & Lei, S. L. (2012). A structural model of residents' intention to participate in ecotourism: The case of a wetland community. *Tourism Management, 33*, 916–925.

Community-based ecotourism managing to fuel community empowerment? An evidence from Malaysian Borneo

Velan Kunjuraman

ABSTRACT

Community-based ecotourism in developing countries is often regarded as a rural community development tool, and its value in Malaysia has been acknowledged by the Malaysian government. This paper presents the case one of the pioneer ecotourism destinations in Sabah, called Bilit village, and its role as the catalyst to rural community development. Using qualitative research methods, this article explored the empowerment indicators gained by the local community from the economic, socio-cultural and environmental aspects, through its active participation in community-based ecotourism. The study discovered that community-based ecotourism helped improve the local community and afforded them an alternative livelihood in their rural destination. However, the sustainability of community-based ecotourism projects in rural destinations is still questionable if the relevant stakeholders fail to consider the community's vision and aspirations. Finally, the study extends an empowerment framework with the novel concept of 'socio-cultural empowerment'.

Introduction

Over the last two decades, community-based ecotourism (CBE) have uncovered several economic, socio-cultural and environmental benefits to relevant nations and their local communities. The benefits delivered by the CBE are manifold (Gan, 2020). Appreciating CBE as a development tool, many countries as depicted in Table 1, have introduced CBE as a rural development project for the local community. CBE contribution to the nation, society and environment is well-acknowledged. For instance, CBE is advocated as a means to acquiring new skills and occupations, job training, business investments, enhancing foreign language proficiency, empowering of local communities, enhancing environmental consciousness, conservation and preservation of natural resources, as well as tourism-driven infrastructural development (Afenyo & Amuquandoh, 2014; Ahmad, 2014; Bhuiyan et al., 2011; Walter, 2011; Hussin, 2008, 2009; Hussin & Mat Som, 2008; Nair & Hamzah, 2015; Rowat & Engelhardt, 2007; Scheyvens, 1999; Stronza & Gordillo, 2008; Stronza & Pêgas, 2008).

Similarly, Lee and Jan (2019), opined that the community-based tourism (CBT)'s main purpose is to enhance economic, socio-cultural, environmental and life satisfaction conditions, which in turn, stimulates the community members to support CBT development. Nevertheless, Gan (2020, p. 2) argues 'there are, undeniably, drawbacks to CBT'. In line with this notion, several negative aspects exist, and that there are 'social, psychological, and political complexities of development initiative impacts of CBE on the local communities' discovered in previous studies. These complexities include women's burdens of domestic labour in tourism (Reimer & Walter, 2013), high number of tourist visitation exposing the community to volatile forces of modernity (Gan, 2020), loss of communal land (Sebele, 2010) consolidation of small-scale agricultural exploitations (Zapata et al., 2011) as well as cultural displacement which lead to immoral activities among the community (Wu et al., 2014). Due to these costs of CBE, the community members believe that CBE had incurred more costs than benefits (Mensah, 2017).

However, CBE's latent as a community development approach is still acknowledgeable (Kunjuraman & Hussin, 2017; Mtapuri & Giampiccoli, 2014) and can be seen in community development plans in many developing countries, including Malaysia (Hussin, 2006). Some recent evidences clearly voiced out such a premise. In Indonesia, for instance, CBE significantly contribute to the trust and collective action among the community members in the village as well as empower them to gain knowledge on economic, social and

Table 1. Case studies of CBE from different countries.

Countries	Authors
Africa	Manyara & Jones, 2007; Ezeuduji & Rid, 2011; Giampiccoli & Kalis, 2012; Isaac & Wuleka, 2012; Afenyo & Amuquandoh, 2014; Burgin & Zama, 2014
Laos	Kim et al., 2014
Thailand	Kontogeorgopoulos, 2005; Kontogeorgopoulos et al., 2014; Pookhao, 2014
New Zealand	Weinberg et al., 2002
Costa Rica	Weinberg et al., 2002
India	Dogra & Gupta, 2012; Sood et al., 2017; Agarwal & Mehra, 2019
America	Zapata et al., 2011
Botswana	Sebele, 2010
Philippines	Okazaki, 2008
Cambodia	Reimer & Walter, 2013
China	Gu & Wong, 2006
Malaysia	Hussin, 2006, 2008, 2009; Hamzah & Khalifah, 2009; Harris, 2009; Ibrahim & Razzaq, 2010; Kayat, 2010; Hussin & Kunjuraman, 2014; Kunjuraman, et al., 2015; Kunjuraman & Hussin, 2017

environmental aspects (Pramanik et al., 2019). In another study by Agarwal and Mehra (2019), CBE homestay programmes are seen as a means to deliver numerous socio-economic benefits. For example, for the local community in Tirthan valley, Himachal Pradesh, India, these benefits come in the form of income earned, number of jobs created, women entrepreneurs involved, reverse migration, development of infrastructure and rise in standard of living. In a similar vein, CBE homestay programmes have offered several economic and social advancement on homestay hosts in Thailand, as documented by Kontogeorgopoulos et al. (2014). The manifold benefits of CBE on local community in economic and social aspects are evident based on previous studies; yet, the challenges are still endured (Kunjuraman & Hussin, 2017; Pusiran & Xiao, 2013). Interestingly, a study by Kunjuraman and Hussin (2017) in CBE homestay programmes in Sabah, Malaysia, provide hopeful insights which point to the fact that successful involvement by the local communities in CBE homestay programmes can indeed lead towards positive economic and social developments. As a result, CBE's roles as an economic and social development tool to the rural communities in developing countries are evident and it is 'therefore, premature to say that CBT is not useful for rural communities and it is unfair to generalise that CBT projects are a failure' (Stone & Stone, 2011, p. 111).

To manage CBE in a sustainable way, local community participation in tourism development is warranted (Murphy, 1985; Telfer & Sharpley, 2008). Murphy's (1985) work had been further extended by Murphy and Murphy (2004) involving recent elements of community participation issues in tourism development with more emphasis on community participation in tourism planning and management of tourism. Similarly, Pawson et al. (2016) are also in agreement that Murphy's work on community is acknowledged in tourism academia, and continuous attempts have been made by tourism scholars to pay more emphasis on community participation in tourism planning and management. In view of this, the current study focuses on community participation in CBE schemes in a developing country context. CBE advocates as a management model to provide opportunities to gain empowerment (Ernawati et al., 2017). The empowerment gains by the local community in tourist destination can be convened in social, economic, and political aspects. Scheyvens (2000, p. 233) noted that 'it has only been in the last few years that empowerment has attracted the interest of those writing on ecotourism'- yet studies related to this area, especially in the context of a developing country, are given scant attention by the tourism scholars with some notable exceptions (Scheyvens, 2000; Tran & Walter, 2014; Kunjuraman & Hussin, 2016). Moreover, such studies are only focused on women without involving the community as a whole. Thus, there is limited understanding on exactly how the local community's continuous participation in CBE schemes empower and improve their livelihood.

Guided by the Longwe's (1995) empowerment framework and Scheyvens' (2000) community empowerment framework in ecotourism, this qualitative study aims to seek underlying factors of empowerment gains by the CBE project participants in the social, economic, and environmental aspects. For this reason, a locally established CBE project in Bilit village, Lower Kinabatangan has been chosen as a case study where such a CBE project was introduced in the early 2000s and has been successfully managed by the local community ever since. It is worth to note that the Lower Kinabatangan is a famous ecotourism destination in East Malaysia among both local and international tourists due to its untouched natural resources. The active local community participation in ecotourism activities in the region provide the local community the opportunity to engage in alternative livelihood activities. The participation has also helped Kinabantangan position itself as an example of a locally-owned community-based ecotourism destination in Malaysia, as well as to other potential similar destinations in the world. Ecotourism in Bilit provides the local community with an opportunity to empower themselves in ecotourism as one of the alternative livelihood activities, where traditional fishing activity is the main economic activity. Due to this reason, CBE in Bilit has been chosen as the research venue to study the level at which CBE empowers the local community economically, socially and environmentally. On the other hand, the rationale behind the adoption of such frameworks is that they are able to provide in-depth insights on empowerment based on the dimensions outlined in the frameworks. Moreover, this study extends the empowerment framework popularised by Scheyvens (2000) with the novel concept of 'socio-cultural empowerment', thus contributing to the current body of knowledge about empowerment in CBE scholarship. The relevance of the findings constitutes to the planning and implementation of CBT as practical contribution of the study.

Theoretical background

Sustainable development goals and community development

Serious consideration paid by the United Nations World Tourism Organization (UNWTO) on community development to achieve sustainable development as recorded in the following statement, caught the researcher's attention (https://www.unwto.org/our-focus):

> 'Tourism should take full account of its current and future economic, social and environmental impacts, addressing the needs of visitors, the industry, the environment and host communities'

The above focus was one of the key motives owned by the UNWTO to achieve sustainable development for the tourism industry, as well as the destination community. In order to realise the sustainable development in the tourism industry, The Sustainable Development Goals (SDGs) was introduced as a blueprint which practised a balanced approach with regards to tourism resources and consumption. The following section is devoted to the SDGs' introduction and its importance to be adopted in this study.

The end of 2015 officially closed the achievement of Millennium Development Goals (MDGs). After the MDGs reached its due date, international professional communities began to formulate a new set of development goals similar to, or even extended from, the MDGs (with inclusive and comprehensive goals) known as SDGs (see Figure 1). The SDGs apply to UN member states which incorporate a set of practical and ambitious goals that seek to, *inter alia*, eradicate extreme poverty, enhance human well-being, strengthen human rights and ensure environmental sustainability (United Nations, 2014 as cited in Elkins et al., 2018). With a joint continuous consultation by relevant stakeholders, the SDGs were finalised and were accompanied by 169 targets and 230 indicators (Elkins et al., 2018). According to Griggs et al. (2013), SDGs should be implemented in a way that benefits all mankind as well as safeguard the existing resources of the Earth. Griggs et al. (2013, p. 305) argued that 'the protection of Earth's life support system and poverty reduction must be the twin priorities for SDGs'. They also argue that 'it is not enough simply to extend MDGs, as some are suggesting, because humans are transforming the planet in ways that could undermine development gains' (Griggs et al., 2013, p. 305).

There is an argument that the revisions of SDGs are quite challenging because they carry the benefits of certain stakeholders' interests or goals such as energy provisions and climate change prevention. However, Griggs et al. (2013) managed to tackle this issue by proposing several sets of goals by combining the MDGs with global environmental targets drawn from science and from existing international agreements. However, such proposed SDGs by Griggs et al. (2013) are still in line with previous MDGs and new SDGs in which the driving principles are retained, updated and extended to 2023. They are reducing poverty and hunger, improving health and well-being, creating sustainable productions and consumption patterns, as well as improving lives and livelihoods.

In the context of tourism and more specifically ecotourism, SDGs have directly and indirectly been linked to tourism where it incorporates the social, economic and environmental aspects. Since tourism has great potential to contribute to the SDGs, relevant policies are still needed, and careful examinations are always a priority. Concerns pertaining sustainability should be inculcated and it is our responsibility to make the visions into reality. Accordingly, the Secretary-General of the United Nations, Ban Ki-moon, 'Everyone has a role in sustainable tourism. Every action counts.' (The Tourism Sector and Sustainable Development Goals Report, 2017, p. 5). Tourism has contributed to human development and has helped countries gain economic benefits. In the context of ecotourism, large numbers of international tourists contribute to the economic development of countries that use ecotourism as a development tool, including Malaysia. Thus, it is evident that ecotourism's contributions to the economic and social development of countries are undeniable and have the potential to grow at a great pace. In this study, the SDGs are relevant to be adopted and evidences from this study could contribute to the global development agenda from the context of a developing country.

Community empowerment through tourism

The concept of 'empowerment' has had a strong link and relationship in community tourism studies since several decades ago (Scheyvens, 1999). Such a concept

Figure 1. The 17 SGDs. Source: https://sustainabledevelopment.un.org/?menu=1300.

has been applied in various fields and is very popular among scholars from political and development studies. For instance, Beeton (2006, p. 88) explained the origin of the empowerment concept by saying that 'the term comes from political science, usually in discussions involving the re-assigning of power to a community whose power has been taken away by force'. In contrast, Lenao and Basupi (2016, p. 54) asserted that 'empowerment has become a very popular concept across various fields of study, particularly those dealing with development and politics'. Moreover, the empowerment concept with conceptual and different definitions have been proposed by many scholars in academia. In line with this statement, Lenao and Basupi (2016) delivered their caution in defining the concept of empowerment. They claimed that they did not have any intention to define the concept, rather that it was 'safe to note how it has been defined by a few users before' (Lenao & Basupi, 2016, p. 54). For Beeton (2006), the concept of empowerment is embedded in power relations in groups. She argued that the concept of empowerment has several issues such as who (active local community) participates, and the fact that the disadvantaged (those who need it most) are left out of the process and women are restricted to low-paying service roles (cooks, cleaners) (Beeton, 2006, p. 92). Thus, empowerment is multidimensional in nature and this study attempts to explore the empowerment gains received by the local community in the study site.

In ecotourism research, there are two analytical models or frameworks available to explore the ecotourism's contribution on local community's livelihoods. They are the Longwe Empowerment Framework (LEF) (Longwe, 1995) and the Community Empowerment Framework (CEF) by Scheyvens (2000). Both frameworks in the past have been successfully applied in community-based ecotourism research as analytical tools to provide interesting insights (Scheyvens, 2000; Tran & Walter, 2014). Table 2 provides the main elements of both frameworks which are potentially applicable in community-based ecotourism research in developing countries.

LEF has been developed with the aim to understand how women's participation in ecotourism and

Table 2. Main elements in LEF and CEF.

Longwe Empowerment Framework (LEF) By Longwe (1995)	Community Empowerment Framework (CEF) By Scheyvens (2000)
Control - equal control in decision-making over factors of production and distribution of benefits	Political - voice in all stages of ecotourism project; broadly representative decision-making structures; overcoming of cultural constraints on participation
Mobilisation/participation - equal participation in decision-making processes related to policymaking, planning and administration	Social - community's sense of cohesion confirmed or strengthened by ecotourism activities; equitable distribution of costs and benefits; income used for community projects; greater respect for women
Conscientization - conscious understanding of gender roles and a gender division of labour, and that these can be changed to be more equitable	
Access - equal access to factors of production (land, labour, credit, education, public services)	Psychological - women at centre of cultural preservation; new skills, increased selfconfidence, self-reliance and pride; motivated to participate in community development
Welfare-equal access to material welfare (food, income, shelter, medical care)	Economic - participate in formal employment; establish own businesses; challenge restrictive social norms; adopt new gender roles; control ecotourism income; increased access to productive resources

Source: Longwe (1995); Scheyvens (2000); Tran and Walter (2014); Kunjuraman & Hussin (2016).

development projects may or may not promote the empowerment as a way to enhance their livelihoods and personal development. There are five components in the framework outlined and the women progress towards betterment from the aspects of these components: welfare, access, conscientization, participation and control (March et al., 1999). The move towards betterment in the women 's lives result from their continuous participation in community-based ecotourism projects over time. Leder (2016) suggests that such a framework is applicable to examine whether the development projects are able to meet the main objective, which is women empowerment. In addition, the framework also emphasized the strong political ideology embedded in the development projects to overcome the gender inequality (March et al., 1999). Interestingly, March et al. (1999) exemplified the division of labour involving three main roles as reproductive, productive and community roles and this division was also popularised by the French philosopher Emile Durkheim's, in his book *The Division of Labour in Society* (or *De la Division du Travail Social*), which debuted in 1893. Tran and Walter (2014) further acknowledge LEF as a transformative model for development projects including ecotourism which 'identifies gaps in gender equity and the potential for change in ecotourism project planning, implementation, outcomes and research' (p. 119).

Secondly, the CEF by Scheyvens (2000) focuses on the gender analysis in ecotourism development in third world countries. The CEF however, does not focus on women *per se*, but women-in-community (Tran & Walter, 2014). There are four dimensions or components involved namely, political, social, psychological and economic as measurement for women empowerment. The applicability of the framework in ecotourism research is evident in the context of developing countries (Tran & Walter, 2014; Kunjuraman & Hussin, 2016). Such evidences further enhance its pertinence in ecotourism research in developing countries. The evidences also provide thought-provoking insights, subsequently enriching the body of knowledge.

Yet, through the critical analysis, this study finds both the LEF & CEF frameworks have their own injunctions. One notable limitation is that both frameworks are devoted to the women group as a unit of analysis. Thus, the frameworks may be notoriously viewed as 'gender oriented' without giving emphasis to the whole local community as a unit of analysis. This debate may disclose further understanding to emphasize the need for involving 'community' as a unit of analysis rather than focusing on women as a single unit of analysis. A community not only comprises of women as a different gender - family values, culture and tolerance are also a part of it. Thus, interesting findings may be provided if such a framework is applied in a community research. The applicability and potential of both frameworks may be limited as they solely focus on only women in ecotourism research. Interestingly, Scheyvens (2002) is optimistic that the CEF may be applied to the analysis of the other group (indigenous people or local communities) and not only women group *per se*. This is due to the fact that all the dimensions in the CEF are experienced by the local communities and it is premature to say that such a framework only focuses on women. Tran and Walter (2014) in a similar vein, call for more comprehensive community study to capture the 'various dimensions of social class, ethnicity, culture, kinship, leadership structures and social context' (p. 129). Thus, with this in mind, this study involves community as a unit of analysis to provide comprehensive findings on empowerment. Secondly, both frameworks overlook analysing socio-cultural issues in CBE projects where ecotourism has been considered as a sustainable development tool for many developing countries which are rich with socio-cultural resources (Cobbinah et al., 2017). The socio-cultural related activities such as meal preparation, dance, local dialect communication and other similar activities should be highlighted in ecotourism projects because they are part of the main source for ecotourism, and are often expected by visitors wanting to experience new culture.

Similarly, LEF has also the potential to be applied in a wider context such as local communities because the main elements may be experienced by anyone in any ecotourism destination. Thus, the current study utilized the LEF and CEF to explore the empowerment gains of CBE on local community in the study site. It is worth to note that, women participation in the CBE scheme in this study is evident and is contextualised as community or ecotourism project participants. As such, both frameworks are relevant to be applied in the CBE scheme in this study, where CEF and LEF empowerment gains in terms of economy, social and environmental aspects are to be explored.

Research context

Bilit Village is located along the Kinabatangan River, between the Sukau (downstream) and Batu Puteh (upstream) villages. In terms of distance, Bilit is reachable about 130 km from Sandakan city, which is an approximately three-hour journey (Figure 2). Now, locally owned private buses (minibus) are provided by the villagers. However, there is no public transportation service to date due to low demand and the fact that the

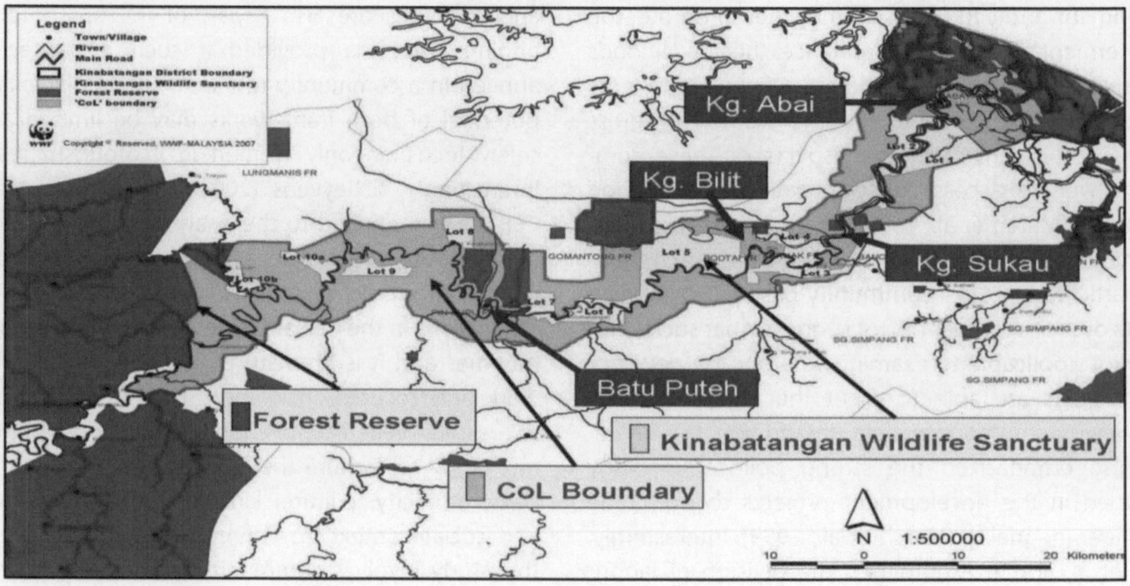

Figure 2. Location of Bilit village (*Kg. Bilit*) in Lower Kinabatangan, Sabah, Malaysia. Source: Fieldwork, 2018.

majority of the locals have their own transport. Alternatively, Bilit is reachable by an approximately 30-minute (25 km) drive from Sukau. The journey may also possibly take only 20 min, due to the improved road conditions. Many locals still practise fishing as their main livelihood and economic activity. Some of them are engaged in small businesses (e.g. grocery shops) and some work at nearby palm oil plantation companies such as IOI, Hap Seng, Bukit Kretam, Sime Darby and Avicess.

In the late nineteenth century, Bilit became a prominent hub for birds' nest harvesting and this activity continues to date. This activity is an important source of income to the locals. Moreover, bird harvesting in the *Gomantong* Caves located in Bilit has also become an attraction for tourists, prompting efforts by the local stakeholders to popularise it. A bird nest harvesters' trail to the *Gomantong* Caves used to be actively travelled in the old days (Vaz & Pyne, 1997). Presently, the *Gomantong* Caves have been visited by both domestic and international visitors to watch the bird nest harvesting, and, if they are lucky, the chance to also watch orangutans and other wildlife animals such as birds and squirrels. Bilit began its first ecotourism project in 2002, when the locals incorporated themselves in the homestay programme introduced by the Ministry of Tourism, Culture and Environment, Sabah (previously known as the Ministry of Tourism, Environment and Technology, Sabah) and World Wildlife Fund (WWF) Malaysia (Hussin, 2006). Similar to Sukau, Bilit offers a variety of ecotourism products through the nature available in the Kinabatangan River.

Bilit has been transformed through ecotourism where now, many lodges, resorts, B&Bs, Jungle Camps and homestays are established to provide multiple types of accommodations to the visitors. In fact, many service providers in Bilit outnumber locally owned accommodations. It was found that many locals are still not interested in being involved in the resort and B&B business as the operational cost is high, preferring to be involved in operating homestays instead as they are relatively cheaper to run. The villagers view homestays as another option for them to be involved in ecotourism-related activities income despite many still mainly depending on fishing. People of Bilit acknowledge that ecotourism has multiple benefits in the economic, socio-cultural and environmental aspects, which is the main focus in this study. The social transformation of local communities which leads to empowerment through ecotourism was explored in the Bilit context where the community has been involved in ecotourism development since 2002, and continue to be involved to date.

Research methodology

There is a concern amongst researchers (Kunjuraman & Hussin, 2017b; Rajaratnam, 2018) where many qualitative studies in tourism often fail to justify the research paradigm. This is a serious issue as a qualitative study should employ an appropriate research paradigm to guide the researchers to achieve study objectives. Accordingly, the current study adopted the constructivist research paradigm, which is also known as the interpretivist research paradigm (Denzin & Lincoln, 2000; Jennings, 2010; Neuman, 2006) in literature aiming at capturing the subjective norms and realities of the informants (Guba & Lincoln, 1994). The field study resulted in a thick and rich description on CBE benefits to the local community, which contributed to the empowerment process in the community level. This, enabled the researcher to adopt an exploratory qualitative research design. A qualitative research study using in-depth interviews and field observation was performed in order to get primary data required for this study. A fieldworker, a local Malaysian from Kota Kinabalu carried out the fieldwork in Bilit village and stayed in the village homestay as a paying guest. Before the actual fieldwork took place, the researcher had identified the gate keeper in the village who is a local homestay coordinator in Bilit village. Being a tourism sociologist, the researcher was given several opportunities to embark on tourism and biodiversity related research in the Lower Kinabatangan region, which enabled him to build rapport with the local community in Bilit village.

In qualitative research, the smallest acceptable sample for an in-depth interview is one and it is suggested that this number can be of great use for increasing one's knowledge (Boddy, 2016; Kaur & Kumar, 2020). To justify the adopted sample size, qualitative scholars should refer to data saturation and homogeneity of the population (Boddy, 2016). A small sample size is enough when the research participants belong to the homogeneous group and data saturation can be achieved after six in depth-interviews (Guest et al., 2006). Accordingly, Kaur and Kumar (2020) carried out a qualitative study involving seven informants which provided meaningful insights and achieved the study's objectives. In this study, a total of sixteen informants (see Table 3) were recruited based on the purposive sampling technique (Sekaran, 2003) to conduct in-depth interviews. In addition, maximum variation sampling in purposive sampling technique was adopted to gain multiple perspectives of individuals about the issue under investigation (Omona, 2013), as well as to show variations in perspectives by informants regarding the issue under studied (Creswell, 2002).

The actual fieldwork was carried out in Bilit village over 6 months, between 2017 and 2019. The key informants of the study are the local community who are the pioneers of CBE projects since it began in 2002.

Table 3. Summary of key informants.

Informant No.	Role category	Total interviews
1–13	**Local community**	**13**
14, 15	**Local government**	**2**
16	**NGO director**	**1**
Total		**16**

Source: Author (Fieldwork, 2018).

The informants are natives of Bilit and are involved in ecotourism activities as an alternative livelihood activity besides engaging as fishermen, small entrepreneurs and oil palm workers. The selection of informants was made on the basis that they are Bilit's pioneers of CBE projects in which they have been actively engaged since 2002. The informants were also selected as they are viewed as being highly capable in empowering themselves from the economic, social and environmental aspects. Additionally, two government officials and a locally operated Non-government organization (NGO) were approached by the researcher to gain additional data required for this study, mainly on the significance of the CBE in the village. The CBE projects' participants in this study are the natives who operate homestay programmes and privately-owned resorts, as well as Bed and Breakfast (B&Bs) in the village. The empowerment issues of CBE on local community betterment were discovered by the researcher and this motivated him to explore them empirically.

All interviews were conducted in the informants' homes and government offices. Letters of concern were distributed to the informants involved and these letters were signed by the informants to signify agreement to be informants in this study. Several interviews were performed twice with the informants, especially the homestay operators and B&B owners, to gain more data related to empowerment issues. All the interviews were done in both *Bahasa Malaysia* (national language of Malaysia) for the CBE project participants and English for the local government officials and NGO director. After the interview session with the CBE project participants, the data were then translated into English by the researcher and finally checked by the English language expert for internal consistency.

Special care was taken by the researcher to ensure parameters for validity in qualitative research known as trustworthiness, which includes four criteria for qualitative inquiry: credibility, transferability, dependability, and confirmability (Lincoln & Guba, 1985). All these criteria were established in this study as shown in Table 4. In addition, the triangulation technique was utilized through the variety of sources and methodological techniques, which warranted the quality of the data. Each interview lasted between one to two hours and was audio-recorded after gaining permission from the informants. Field observation was performed during the fieldwork to observe the CBE operations, physical environment, infrastructural development in the village, and ecotourism activities. All observations were captured in the form of fieldnotes and photographs which add value to the main qualitative interview data. In order to explore the empowerment

Table 4. Four criteria for qualitative inquiry adopted in this study.

Types of trustworthiness	Adopted by the researcher
Credibility	Credibility was established by the researcher using techniques of prolonged engagement, continuous observation and member checks. The researcher is a local Malaysian who has been engaged in ecotourism, conservation and community development projects for more than five years in the study sites. The interview and observation data were viewed by the study informants as a form of the researcher's interpretation on data collected.
Transferability	Transferability was enhanced in this study through the purposive sampling used.
Dependability	Dependability was accounted in this study through the development of details research plan, which includes the research process, audit trails of the transcripts and discussion of the project over time between two researchers.
Confirmability	Confirmability was enhanced through the research audit process. In this process, the researcher was aware of the variety of explanations used in discussing the phenomenon being studied.

indicators gained by the informants through their active engagement in CBE projects in Bilit, the six stages in thematic analysis techniques were applied. These six stages are: familiarising the data, generating initial codes, searching for themes, reviewing themes, defining and naming themes and producing the report, enabled the researcher to identify several themes based on the study objectives (Sood, et al. 2017; Kunjuraman & Hussin, 2017). In accordance with the principle of writing up qualitative research, the informants' quotes illustrating responses were selected 'through judicious selection of statements emblematic of an issue being illustrated and often indicative of the voices of other participants who felt similarly' (Sood et al., 2017, p. 338).

Research findings

The findings of the study reveal that the dimensions embedded in both LEF and CEF have been captured. Those identified dimensions were found in the findings based on economic, socio-cultural and environmental aspects as CBE impacts on the local community, and these are presented in the next section. The essence of the CBE impacts is described in the text, illustrated by the relevant quotes, and triangulated to present rich information. As shown in Table 5, CBE project participants are mainly involved in homestay programmes which are their primary and secondary form of employment. Fishing is also performed by some informants to gain additional income. Interestingly, there is an informant, who is a local, who successfully set up a lodge and was able to venture into the ecotourism business.

Table 5. Background information of CBE project participants.

Informant No.	Gender	Age	Main job	Secondary job
1	M	46	Homestay	Fishing
2	M	56	Homestay	Fishing
3	F	56	Homestay	Fishing
4	F	39	Homestay	Fishing
5	M	30	Homestay	Fishing
6	F	53	Homestay	Fishing
7	F	42	Homestay	Fishing -
8	M	44	Homestay	Fishing
9	F	59	Homestay	Fishing
10	F	47	Homestay	Fishing
11	F	32	Homestay	Fishing
12	F	46	Lodge servant	Homestay
13	M	47	Teacher	Lodge owner, contractor & palm oil farmer

Source: Author (Fieldwork, 2018).

He may be considered as a local 'champion', who is a role model to other CBE project participants to venture into ecotourism business. Out of the thirteen informants, eight are women. These women manage homestays which exemplify their capabilities. The informants are between the ages of thirty and fifty-nine. It was observed that the young people in the community who are below thirty, also participated in CBE projects as helpers for their parents. They mainly helped with guiding the tourists, preparing food and cleaning the homestays.

CBE economic impacts and empowerment

CBE offers green job opportunities

The social transformation of local communities through ecotourism can be seen in the economic aspect, especially transformations of jobs from fishing to homestay programmes and lodge entrepreneurs which are considered as green job opportunities in the rural areas. Bilit is blessed with untouched natural resources, namely flora and fauna, which have great potential for ecotourism development. By being involved in the homestay programmes, the informants are aware that Bilit's natural resources can be promoted as an ecotourism product to the tourists and at the same time, generate income. The establishment of ecotourism in Bilit provide opportunities for the local community to be engaged in more sustainable livelihood activities such as the homestay programme to enhance their lifestyle. There was a general agreement among the informants that ecotourism delivered a positive social transformation to the local community in the employment sector. Through the access to ecotourism, it is evident that some in the local community in Bilit had transformed their employment from traditional fishing to homestay operators. For instance, an informant commented:

> 'Here [Bilit], the majority are working as fisherman, and some of them are involved in farming. Whenever a homestay exists, it is considered as another economic activity and alternative job opportunity for the interested villagers.' (Inf. No. 5, homestay operator at Bilit)

The informants who were involved in the homestay programmes had gone through a social transformation in the context of employment whereby ecotourism gives them the chance to improve their economic welfare, which is not provided by fishing and subsistence farming. In addition, the informants also agreed that the homestay programme not only provides an alternative green job opportunity to the local people, but also reduce their dependence on traditional fishing activities.

Income generation

Fishing and subsistence farming have been the main livelihood activities among the local community in Bilit for decades. However, ecotourism has provided the selected local community (those interested in ecotourism) a chance to transform their socio-economic welfare. Generally, the income Malaysian Ringgit (MYR) from employment are diverse in Bilit. The informants stated that the side income was directly and indirectly received through the participation in ecotourism activities such as homestay programmes, B&Bs and resorts, becoming tour guides, being involved in forest restoration programmes, handicraft making, providing boat services, bus services and taking part in local cultural performances in homestay programmes in Bilit. For instance, a teacher from Bilit has successfully established a lodge known as the Kinabatangan Sunshine Lodge due to his motivation to diversify his income. Being a teacher, he felt that the ecotourism business brought him economic social transformation through increased sustainable income. He noted that the income obtained from operating the lodge is significantly much more than what he makes from teaching. He commented:

> 'The lodge, which is based on ecotourism, increased my monthly income from MYR300 to MYR120 K. This motivated me to be involved more actively in ecotourism projects and I will continue to be actively involved.' (Inf. No. 13, Lodge owner at Bilit)

It is clear that ecotourism in Bilit had diversified this informant's income through his involvement in other economic activities like running the lodge which can be viewed as a new job opportunity to the local community in Bilit. Such engagement in ecotourism projects by the informant motivated him economically to find ways to boost his business. However, based on the research findings, the study indicates that the income from ecotourism received by the local community, especially those who were from the homestay programmes (as opposed to those who run lodges), is relatively small-and this leads to disempowerment. Nevertheless, the informants of the study (the homestay operators) reported that they were satisfied and felt grateful for the ecotourism revenue. In addition, the study found that the villagers perceived working in ecotourism as relatively easier in comparison to other economic activities such as fishing, farming and subsistence activities. It was discovered that the ecotourism activities in Bilit provided sufficient income to the homestay operators if the number of tourists increases every month. A number of informants reported that the minimum income received from the homestay programme was MYR300, and this was mainly used for daily household activities. This income is still insufficient to fulfil the daily needs of the households and contributed to an economic disempowerment; but the informants are grateful to receive such income. Although the homestay programme provides a small income to the homestay operators at Bilit, the contribution towards the enhancement of community bonding, cultural preservation and conservation awareness was acknowledged by the informants of the study.

CBE socio-cultural impacts and empowerment

Strong social capital

Developing social capital within and outside the community will bring many positive benefits to the local community in the tourists' destination. Similarly, social capital among the informants is one of the community's strengths to secure future livelihood. It was revealed by the informants that before ecotourism existed, the local community did not have any contact with outsiders, especially issues related to development. The local community paid less attention to the cooperation and its importance to human development. There was a time when the local community did not care to work together with outsiders due to little exposure and awareness at the community level. This was explained by the homestay coordinator of Bilit as follows:

> 'The local community is now well-connected due to ecotourism and this was not seen in the past decades. They were busy with their own lives to generate income for the family. After the introduction of the homestay programme in the early 2000s, they received exposure to be involved in tourism activities.' (Inf. No. 1, homestay operator at Bilit)

Due to community's earlier attitude of not working together with outsiders, Bilit was not able to diversify its economy and failed to enhance the community's livelihood. As a remedy, the stakeholder's participation through Non-government organizations (NGOs) was first established in the Lower Kinabatangan region in the 1980s to develop and introduce conservation efforts. In Bilit, WWF Malaysia was the first NGO that managed to establish the first homestay programme in 2002 with support from interested households. With the establishment of the homestay programme, CBE became an influence on the local community - they were now quick to embrace the importance of the external factors in contributing to the village ecotourism development. The collaboration of stakeholders in ecotourism development to embrace social networks among the local community still prevails. Without a social network, establishing ecotourism in Bilit would not have been possible. Social transformations and empowerment of the local community would also then be missing.

Local Orang Sungai culture as ecotourism product

The informants of Bilit feel proud of their local culture, and even more so after the introduction of ecotourism in the village. This is because ecotourism-based activities such as homestay programmes and lodges emphasise the importance of the local culture through cultural performances to the guests. The homestay programme is the main platform for the informants to promote their culture of *orang sungai* through cultural performances and a welcome dance for the tourists. The commodification of the local culture through ecotourism was consulted by the local community and received consent. Evidently, according to one of the informants:

> 'The local traditional activities like dancing and cooking have been promoted as homestay products since their introduction in the village. After the local community agreed, we wanted to promote them through the homestays.' (Inf. No. 3, homestay operator at Bilit)

The local culture is preserved by the local community through decades of practise. There was an agreement among the informants (homestay operators, government officials and NGO director) that the commodification of the *orang sungai* culture can be promoted as an ecotourism product in the homestay programme to receive potential economic benefits. The consultation was properly done to discuss the extent to which the local culture should be promoted through the homestay programme. Amongst the points discussed was the benefits the local community stand to gain, by promoting their culture. It is clear in the interviews with the informants that the representation of the *orang sungai* culture can be manifested through traditional costumes, food, dances and singing, all of which are often demonstrated and performed by the informants regardless of gender, age and social status. Through the homestay programme, the informants perceived positive impacts such as the preservation of the local culture and the feeling of pride towards their own *orang sungai* culture. It is also worth noting that the informants did not realise the importance of the local culture until its promotion through ecotourism. They revealed that the *orang sungai* culture was not exposed to outsiders as they did not consider sharing it with others as good practice. After the introduction of ecotourism, the informants came to realise the potential of the local culture as a resource for ecotourism. However, the informants also realized the authenticity of the *orang sungai* culture should be preserved and not overly commercialised. For this reason, they try to minimise exhibiting and conducting all activities related to their local culture to only once in a week, for stays that are at least a week long.

Women empowerment

The establishment of ecotourism in Bilit showed equitable gender involvement in ecotourism-related activities. Both men and women were equally and justly acknowledged for their contributions in ecotourism development as an output of conscientization. Traditionally, the *orang sungai* culture emphasised the patriarchal system where the men are the dominant group in the society. It is increasingly challenging for women to break the system which has been inherited for centuries. However, this system has slowly been broken down due to ecotourism and women are empowered socially and politically through their engagement in the homestay programmes. Presently, women who are engaged in the homestay programmes are more confident in managing the homestay and in communicating with outsiders compared to those who are not involved in the homestay programmes. The observation data revealed that women in the homestay programmes learn management skills such as organising cultural events for tourists, preparing daily meals, making handicraft, managing nurseries and developing proper communication skills. Such social benefits received by the women are recognised as social transformation impacts delivered by ecotourism in Bilit.

There are some evidences in this study that show how women are socially empowered through their involvement in the homestay programmes at Bilit. The evidences are based on the social aspect such as cooperation, knowledge on ICT, and management skills. The female informants (Inf 3, 6 & 7) in the study documented their positive social transformation by engaging in the homestay programmes. For instance, the women homestay operators learnt management skills in operating the homestay programmes from the registration until the end of stay of the tourists. Another informant described her role in the homestay programme where she had learnt to market her homestay programme to outsiders through the internet. Such a skill was cultivated by the homestay programme after her participation and she credited ecotourism for the opportunity for empowerment.

CBE environmental impacts and empowerment

Increasing awareness on environmental protection

The informants of the study clearly shared their views that the natural environment of Bilit has been receiving more attention since the 1990s, when ecotourism slowly started emerging as a remedy for illegal activities such as deforestation and illegal logging, which were rampant in the Lower Kinabatangan region. As shared by the homestay operator:

> 'Many years ago, logging was rampant in this area and it slowly stopped after the local government announced Lower Kinabatangan as a 'conservation area'. The local community still depends on the natural resources for their daily survival such as fishing and farming.' (Inf. No. 1, homestay operator at Bilit)

The environment of Bilit was challenged by many parties including the local community who depend on the resources to sustain their livelihood. Outsiders aimed the Lower Kinabatangan region as a 'golden land' where the commercial value can be seen through the abundance of natural resources. The empty land was converted into agricultural land and was home for many plantation companies which mushroomed in the early 1980s, in the Lower Kinabatangan region, including Bilit. Due to such attempts, the environment of Bilit was seriously affected and became a threat to the nature. A noble initiative to safeguard the nature was then undertaken by the WWF, with support from the state government. This initiative resulted in a conservation centre which is known as the Lower Kinabatangan Wildlife Sanctuary (LKWS).

Initially, the LKWS was not fully accepted by the local community as it restricted the daily livelihood activities of the local community who relied on the natural products of the forest such as wood and marine resources. However, continuous initiatives by the state government and several NGOs successfully transformed the mind-set of the local community as psychological empowerment to love nature and prevent it from depleting. Obviously, ecotourism became a development tool which transformed the local community into becoming aware of the importance of the nature of Bilit. The importance of the environment in Bilit became a priority in ecotourism activities which is in line with one of the ecotourism principles. The lack of awareness on the environment was limited before ecotourism was introduced in the village. The informants acknowledge that ecotourism has successfully transformed them into having an eco-friendly mind-set. On the other hand, the informants are also aware of the fact that an influx of tourists, or crowding, would bring negative impacts to the environment. However, to date such a phenomenon has not yet occurred, because the homestay programmes are receiving smaller numbers of tourists every month (below fifty). This may limit the environmental costs.

Discussion and conclusion

With the focus on how CBE schemes impact the community empowerment, this study reveals that local community involved in CBE projects (homestay programme & B&B) are able to transform their livelihoods from the economic, socio-cultural and environmental aspects. There is a strong agreement among the CBE project participants that ecotourism exerts positive influences on their livelihoods, in terms of providing green job opportunities, generating income, strong social capital, preservation of local culture, women empowerment and awareness on environmental protection. Such benefits identified in this study further increased the value of ecotourism as a livelihood activity in rural destinations as well as enhance support for tourism development (Lee & Jan, 2020). In terms of the economic aspect, CBE's benefits can be discovered through the job opportunities in the ecotourism scheme where previously, the community was largely made up of fishers and farmers. Such phenomenon observed in this study managed to reduce dependency on marine fishing resources (Porter et al., 2015). Such transformation from fishing to ecotourism entrepreneurs (homestay operators and B&B owners), can be taken as economic empowerment (Scheyvens, 2000). It is evidently reported that the transformation in rural destinations like Bilit enhanced the potential of ecotourism as a development tool and this is also observed in other CBE destinations in Malaysia (Hussin et al., 2015).

With focus to sustainable community development, the study reveals that socio-cultural impacts of CBE on community empowerment outnumbering economic and environmental impacts in a developing country are hardly found in the previous literature. For instance, strong social capital within and outside the village, resulting from participation in CBE has given the local community an opportunity to 'accept' and 'welcome' an alternative livelihood opportunity. The acceptance of CBE as an alternative livelihood activity by the local community in Malaysia was revealed in this study which values the contribution of CBE as an empowerment tool. Limited studies on local community participation in CBE activities in Malaysia fail to acknowledge ecotourism as a community empowerment tool. Consequently, by having a case study from Malaysia, the current study adds value to CBE having the potential to enhance local community's social well-being. The findings of this study also add new perspectives to the current body of knowledge on CBE and community empowerment literature. Moreover, incorporating tourism along with traditional activities like fishing may enhance livelihood sustainability and this study observed findings similar to that as reported by Su et al. (2019) in China. However, there is evidence that fishing is no longer a popular livelihood activity after ecotourism was introduced as an alternative livelihood.

This study discovered that the local community has been empowered by ecotourism in line with CEF's political, social, psychological and economic dimensions (Scheyvens, 2000). LEF dimensions of welfare, access, conscientization, participation and control were also captured in this study. Most importantly, women have improved their social status by participating in CBE projects. Women within the community have been empowered by the CBE (Acharya & Halpenny, 2013) as equal decision makers, and active participants who are also given access to additional income. This study is in line with a previous study by Irandu and Shah (2014) where the women in Isecheno Women's Conservation Group in Kenya managed to enhance their management skills by participating in the project with 100 per cent participation rate. However, this study did not support the findings of the study conducted in Cambodia, by Reimer and Walter (2013), where women are intensified with burdens of domestic and tourism jobs yet, played a vital role to the ecotourism development (Scheyvens, 2000). On the other hand, ecotourism principles as a natural conservation tool (Nair & Hamzah, 2015) were evident in this study where the local community is aware of the importance of nature to the ecotourism development.

The current study offers some significant theoretical contribution. First, through the critical analysis of previous ecotourism scholarship, the current study contributes to the ongoing discussion about the achievement and short comings of the CBE in developing countries. Towards this end, the research findings positively influence the fact that CBE has empowered the rural communities in developing countries, as well as being acknowledged as an alternative livelihood approach (Tao & Wall, 2009). Second, the research findings added new insights to the CBE and empowerment to the body of knowledge where the local culture has an independent dimension in the empowerment framework, which was missing. This study observed that local culture has been commercialised through ecotourism by the CBE projects especially homestay programmes. The preservation of local culture should be highlighted because it is considered as a part of prime ecotourism products, second to nature-based products like wildlife/animal watching. The promotion of the local culture which is embedded in social life through ecotourism, provided the local community with a sense of pride and established in what is known as socio-cultural empowerment gains. To enrich the literature, this particular interesting finding needs to be highlighted as an individual empowerment indicator in CEF as 'socio-cultural empowerment' rather than having named it as psychological empowerment *per se*. The psychological empowerment indicator by Scheyvens (2000) notes women as *the group which is central to cultural preservation*; however, the current study argued that the community as a whole, is the one that bears the responsibility to preserve their local culture. In the case of CBE in Bilit, socio-cultural empowerment was clearly discovered by having the *orang sungai* culture as the main source of ecotourism products for the visitors. Thus, this study extends the CEF with the new concept of 'socio-cultural empowerment' (see Figure 3).

In addition, this study has provided new insights to the relevance of LEF and CEF in ecotourism research which were commonly applied on women in ecotourism and development studies. As a result the researcher consider them as 'gender oriented' frameworks. This particular study is the pioneer in utilizing the LEF and CEF in CBE studies which focus on the local community as a unit of analysis, as compared to other studies which give emphasis to women as a single unit of analysis (Scheyvens, 2000; Tran & Walter, 2014; Kunjuraman & Hussin, 2016). The use of LEF and CEF in CBE studies help to identify the empowerment indicators experienced by the local community in developing countries, and the current study represents a case study from Sabah, Malaysian Borneo. Similar studies utilizing LEF and CEF to explore the empowerment gains received by the local community of other ecotourism destinations, are needed to confirm the utility of such frameworks. On the other hand, this study positively influences several SDG goals, namely one, four, five and sixteen (see Table 6). By having a case study from a developing country like Malaysia, it is proven that ecotourism has

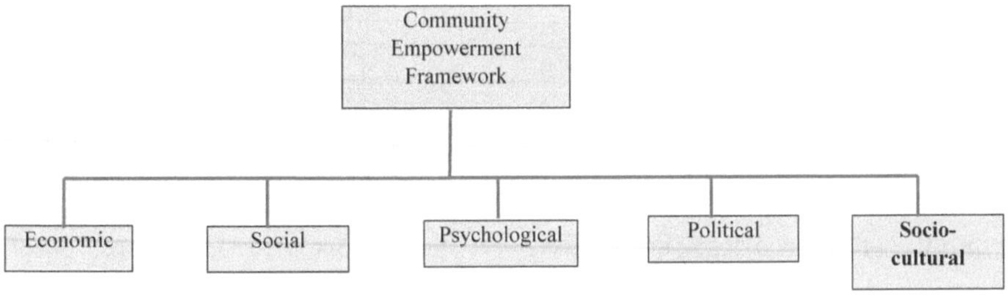

Figure 3. Revised Community Empowerment Framework. Source: Author's plot, 2020.

been acknowledged as a development tool for nature protection as well as for rural community development.

This study alerted that the sustainability of the ecotourism in the study site is questionable in delivering benefits which are perceived by the local community. This is due to the fact that, the CBE project participants, especially homestay hosts, gained limited economic benefits, and this subsequently led to economic disempowerment (Scheyvens, 1999). Due to the limited number of tourists arriving and staying at homestays under the homestay programme, income generation is limited. However, the homestay hosts are still optimistic and grateful for the limited income, as it still helps pay for daily household expenses. Their motivation to be involved in ecotourism may enhance the sustainability in the long run with the support from the relevant stakeholders. Thus, the current study proposes several managerial implications to support the survival of CBE projects in Bilit, in order for them to continue flourishing as a community development tool. For instance, the government and NGOs working in development projects, including ecotourism, should introduce mentorship programmes. These programmes, which may include talks that raise motivation, awareness and skills, can help the local community address fundamental issues and venture into ecotourism scheme on a higher level. There are several success stories in the region where several households managed to transform their homestay to B&B and resorts with better facilities and services. Access to the above-mentioned information is required to sustain the ecotourism operation in the village. To encourage the community to participate in tourism and be empowered socially and economically, access to information is necessary (Cole, 2006).

Table 6. SDGs application in Bilit CBE projects.

SDGs points	Application of the agenda in Bilit village
SDG 1-End poverty in all its forms	CBE as a source of local community development, and capable to provide new job opportunities and income generation
SDG 4-Ensure inclusive and equitable quality education and promote lifelong learning for all	Education and trainings concerning environmental awareness through capacity building by stakeholders (NGOs) provided to the villagers
SDG 5-Achieve gender quality and empower all women	Empowerment of women through active participation and income generation through homestay programme and B&Bs
SDG 16-Promote peaceful and inclusive societies, provide access to justice for all and build inclusive institutions	Promotion of tolerance and multicultural understanding as a result of interaction between local community and tourists

Source: Author (Fieldwork, 2018).

Reflecting on the title of this study, the findings indicate that CBE in the Bilit village has fuelled empowerment among the local community who are actively involved, thus, further empowering them in economic, socio-cultural and environmental aspects. The empowerment gained by the local community is due to their continuous participation in CBE, since it began in 2002. The current study has the limitation in its scope of exploration. For instance, the current study only focuses on community empowerment issues in CBE operation without emphasizing the challenges faced by the local community. Throughout the process to gain empowerment, the local community might face some challenges to sustain the ecotourism. Thus, this study recommends more studies to explore such issues. Moreover, future studies may also consider the reasons for non-participation in ecotourism (Sood et al., 2017) by the local population in a similar ecotourism destination. This is because this study is limited to only those who choose to cooperate and be actively involved in ecotourism projects. It is suggested that future research may explore the views of non-participants', in order to understand the impacts of CBE to the whole community in the study site. More empirical studies on ecotourism contribution as a development tool in similar ecotourism destinations may provide interesting findings, as well as contribute to the field of community-based ecotourism and development.

Acknowledgements

The research was supported by the Universiti Malaysia Sabah and Universiti Malaysia Kelantan. A special thanks to the all the informants involved in this study.

Disclosure statement

No potential conflict of interest was reported by the author(s).

ORCID

Velan Kunjuraman http://orcid.org/0000-0002-5616-4712

References

Acharya, B. P., & Halpenny, E. A. (2013). Homestays as an alternative tourism product for sustainable community development: A case study of women-managed tourism product in rural Nepal. *Tourism Planning and Development*, *10*(4), 367–387. https://doi.org/10.1080/21568316.2013.779313

Afenyo, E. A., & Amuquandoh, F. E. (2014). Who benefits from community-based ecotourism development? Insights from Tafi Atome, Ghana. *Tourism Planning & Development*, *11*(2), 179–190.

Agarwal, S., & Mehra, S. (2019). *Socioeconomic contributions of homestays: A case of Tirthan valley in Himachal Pradesh (India). TISC - Tourism International Scientific Conference Vrnjačka Banja*, *4*(1), 183–201. Retrieved from http://www.tisc.rs/proceedings/index.php/hitmc/article/view/251

Ahmad, A. (2014). Ecotourism in Brunei Darussalam: A qualitative evaluation of its sustainability. *Journal of Environment and Human*, *1*(2), 1–10.

Beeton, S. (2006). *Community development through tourism*. Landlink Press.

Bhuiyan, M. A. H., Siwar, C., Ismail, S. M., & Islam, R. (2011). The role of government for ecotourism development: Focusing on East Coast economic region. *Journal of Social Science*, *7*, 557–564. https://doi.org/10.3844/jssp.2011.557.564

Boddy, C. (2016). Sample size for qualitative research. *Qualitative Market Research*, *19*(4), 426–432.

Burgin, S., & Zama, E. F. (2014). Community-Based tourism - option for forest-Dependent communities in 1A IUCN Protected areas? Cameroon case study. *SHS Web of Conferences*, *12*, 1–10, https://doi.org/10.1051/shsconf/201412010

Cobbinah, P. B., Amenuvor, D., Black, R., & Peprah, C. (2017). Ecotourism in the Kakum conservation area, Ghana: Local politics, practice and outcome. *Journal of Outdoor Recreation and Tourism*, *20*, 34–44.

Cole, S. (2006). Information and empowerment: The keys to achieving sustainable tourism. *Journal of Sustainable Tourism*, *14*(6), 629–644.

Creswell, J. W. (2002). *Educational research: Planning, conducting, and evaluating qualitative and quantitative research*. Pearson Education.

Denzin, N. K., & Lincoln, Y. S. (2000). Introduction: The discipline and practice of qualitative research. In N. Denzin, & Y. Lincoln (Eds.), *Handbook of qualitative research* (pp. 1–28). Sage.

Dogra, R., & Gupta, A. (2012). Barriers to community participation in tourism development: Empirical evidence from a rural destination. *South Asian Journal of Tourism and Heritage (SAJTH)*, *5*(1), 129–141.

Elkins, M., Feeny, S., & Prentice, D. (2018). Are poverty reduction Strategy Papers associated with reductions in poverty and improvements in wellbeing? *The Journal of Development Studies*, *54*(2), 377–393. https://doi.org/10.1080/00220388.2017.1299140

Ernawati, N. M., Sanders, D., & Dowling, R. (2017). Host–guest Orientations of community-based tourism products: A case study in Bali. *Indonesia. International Journal of Tourism Research*, *19*(3), 1–16. https://doi.org/10.1002/jtr.2119

Ezeuduji, I. O., & Rid, W. (2011). Rural tourism offer and local community participation in The Gambia. *TOURISMOS*, *6*(2), 187–211.

Gan, J. E. (2020). Uncovering the environmental and social Conflicts behind Residents' Perception of CBT: A case of Perak, Malaysia. *Tourism Planning & Development*, 1–19. https://doi.org/10.1080/21568316.2020.1749121

Giampiccoli, A., & Kalis, J. H. (2012). Community-based tourism and local culture: The case of the amaMpondo. *PASOS*, *10*(1), 173–188. https://doi.org/10.25145/j.pasos.2012.10.017

Griggs, D., Stafford-Smith, M., Gaffney, O., Rockström, J., Ohman, M. C., Shyamsundar, P., Steffen, W., Glaser, G., Kanie, N., & Noble, I. (2013). Policy: Sustainable development goals for people and planet. *Nature*, *495*, 305–307. https://doi.org/10.1038/495305a

Gu, M., & Wong, P. P. (2006). Residents' perception of tourism impacts: A case study of homestay operators in Dachangshan Dao, North-East China. *Tourism Geographies*, *8*(3), 253–273. https://doi.org/10.1080/14616680600765222

Guba, E. G., & Lincoln, Y. S. (1994). Competing paradigms in qualitative research. *Handbook of Qualitative Research*, *2*, 163–194.

Guest, G., Bunce, A., & Johnson, L. (2006). How many interviews are enough? An experiment with data saturation and variability. *Field Methods*, *18*(1), 59–82.

Hamzah, A., & Khalifah, Z. (2009). *Handbook on community-based tourism: How to develop and sustain CBT*. Asia-Pacific Economic Cooperation.

Harris, R. W. (2009). Tourism in Bario, Sarawak, Malaysia: A case study of Pro-poor community-based tourism Integrated into community development. *Asia Pacific Journal of Tourism Research*, *14*(2), 125–135.

Hussin, R. (2006). *Ecotourism development and local community participation: Case studies of Batu Puteh and Sukau village in Lower Kinabatangan area of Sabah, Malaysia* (Unpublished doctoral dissertation). United Kingdom: University of Glasgow.

Hussin, R. (2008). Ecotourism and community participation in the homestay programme of Sukau village: Long-term or limited benefits? *Jurnal Sarjana*, *23*, 72–86.

Hussin, R. (2009). Ecotourism, local community and "partners for Wetlands" in the Lower Kinabatangan area of Sabah: Managing conservation and Conflicts? *Borneo Research Journal, 3*, 57–79.

Hussin, R., & Kunjuraman, V. (2014). Pelancongan mapan berasaskan komuniti (CBT) melalui program homestay di Sabah, Malaysia (*sustainable community-based tourism (CBT) through homestay programme in Sabah, East Malaysia*). *Geografia: Malaysian Journal of Society and Space, 10*(3), 160–174.

Hussin, R., Kunjuraman, V., & Weirowski, F. (2015). Work transformation from fisherman to homestay tourism entrepreneur: A study in Mantanani Island, Kota Belud, Sabah, East Malaysia. *Jurnal Kemanusiaan, 24*(1), 1–15.

Hussin, R., & Mat Som, A. P. (2008). Ecotourism, conservation programme and local community participation: Conflict of interests in Sukau village of Sabah. *Sosiohumanika: Jurnal Pendidikan Sains Sosial dan Kemanusiaan, 1*(1), 115–140.

Ibrahim, Y., & Razzaq, A. R. A. (2010). Homestay Program and rural community development in Malaysia. *Journal of Ritsumeikan Social Sciences and Humanities, 2*, 7–24.

Irandu, E., & Shah, P. (2014). The role of ecotourism in promoting women empowerment and community development: Some reflections from Kenya. *Tourism Hospitality Management, 2*, 245–259. https://doi.org/ 10.17265/2328-2169/2014.06.002.

Isaac, K. M., & Wuleka, C. J. (2012). Community-Based ecotourism and livelihood enhancement in Sirigu, Ghana. *International Journal of Humanities and Social Sciences, 2* (18), 97–108.

Jennings, G. (2010). *Tourism research* (2nd ed.). Wiley.

Kaur, K., & Kumar, P. (2020). Social media usage in Indian beauty and wellness industry: A qualitative study. *The TQM Journal*, DOI 10.1108/TQM-09-2019-0216

Kayat, K. (2010). The nature of cultural contribution of a community-based homestay programme. *TOURISMOS, 5*(2), 145–159.

Kim, S., Park, E., & Phandanouvong, T. (2014). Barriers to local residents' participation in community- based tourism: Lessons from Houay Kaeng village in Laos. *SHS Web of Conferences, 12*, 1–8.

Kontogeorgopoulos, N. (2005). Community-Based ecotourism in Phuket and Ao Phangnga, Thailand: Partial Victories and Bittersweet Remedies. *Journal of Sustainable Tourism, 13*(1), 4–23.

Kontogeorgopoulos, N., Churyen, A., & Duangsaeng, V. (2014). Success factors in community-based tourism in Thailand: The role of luck, external support, and local leadership. *Tourism Planning & Development, 11*(1), 106–124.

Kunjuraman, V., & Hussin, R. (2016). Women Participation in Ecotourism Development: Are They Empowered? *World Applied Sciences Journal, 34*(12), 1652–1658.

Kunjuraman, V., & Hussin, R. (2017). Challenges of community-based homestay programme in Sabah, Malaysia: Hopeful or Hopeless? *Tourism Management Perspectives, 21*, 1–9.

Kunjuraman, V., & Hussin, R. (2017b). Social transformations of rural communities through ecotourism: A systematic review. *Journal of Tourism, Hospitality, & Culinary Arts, 9*(3), 54–66.

Kunjuraman, V., Hussin, R., & Yasir, S. M. (2015). Challenges of seaweed communitybased homestay programme in Sabah, Malaysian Borneo. *Proceedings of the Hospitality & Tourism Conference 2015*, Melaka, Malaysia, 2–3 November 2015. London: Taylor & Francis Group.

Leder, S. (2016). *Linking women's empowerment and their resilience: Literature review.* Technical Report. DOI: 10.13140/RG.2.1.3395.0809.

Lee, T. H., & Jan, F. H. (2019). Can community-based tourism contribute to sustainable development? Evidence from residents' perceptions of the sustainability. *Tourism Management, 70*, 368–380. https://doi.org/10.1016/j.tourman.2018.09.003

Lenao, M., & Basupi, B. (2016). Ecotourism development and female empowerment in Botswana: A review. *Tourism Management Perspectives, 18*, 51–58.

Lincoln, Y. S., & Guba, E. G. (1985). *Naturalistic inquiry*. Newbury Park, CA: Sage Publications.

Longwe, S. (1995). Gender awareness: The missing element in the third world development program. In C. March, & T. Wallace (Eds.), *Changing perception: New writings on gender and development*. Oxfam.

Manyara, G., & Jones, E. (2007). Community-based tourism enterprises development in Kenya: An exploration of their potential as avenues of poverty reduction. *Journal of Sustainable Tourism, 15*(6), 628–644.

March, C., Smyth, I., & Mukhopadhyay, M. (1999). *A guide to gender-analysis frameworks*. Oxfam Publishing.

Mensah, I. (2017). Benefits and challenges of community-based ecotourism in Park-Fringe communities: The case of *Mesomagor* of Kakum national Park, Ghana. *Tourism Review International, 21*, 81–98.

Mtapuri, O., & Giampiccoli, A. (2014). Towards a comprehensive model of community-based tourism development. *South African Geographical Journal*. https://doi.org/10.1080/03736245.2014.977813.

Murphy, P. E. (1985). *Tourism: A community approach*. Routledge.

Murphy, P. E., & Murphy, A. E. (2004). *Strategic management for tourism communities: Bridging the gaps*. Aspects of Tourism Series Channel view Publications.

Nair, V., & Hamzah, A. (2015). Successful community-based tourism approaches for rural destinations– the Asia Pacific experience. *Worldwide Hospitality and Tourism Themes, 7* (5), 429–439. https://doi.org/10.1108/WHATT-06-2015-0023

Neuman, W. L. (2006). *Social research methods qualitative and quantitative approaches* (6th ed.). Pearson Education.

Okazaki, E. (2008). A community-based tourism model: Its Conception and Use. *Journal of Sustainable Tourism, 16*(5), 511–529.

Omona, J. (2013). Sampling in qualitative research: Improving the quality of research outcomes in higher education. *Makerere Journal of Higher Education, 4*(2), 169–185.

Pawson, S., D'Arcy, P., & Richardson, S. (2016). The value of community-based tourism in Banteay Chhmar, Cambodia. *Tourism Geographies, 19*(3), 378–397.

Pookhao, N. (2014). Community-based ecotourism: The transformation of local community. *SHS Web of Conferences, 12*, 1–8.

Porter, B. A., Orams, M. B., & Luck, M. (2015). Surf-riding tourism in coastal fishing communities: A comparative case study of two projects from the Philippines. *Ocean & Coastal Management, 116*, 169–176.

Pramanik, P. D., Ingkadijaya, R., & Achmadi, M. (2019). The role of social capital in community-based tourism. *Journal of Indonesian Tourism and Development Studies, 7*(2), 62–73. https://doi.org/10.21776/ub.jitode.2019.007.02.02

Pusiran, A. K., & Xiao, H. (2013). Challenges and community development: A case study of homestay in Malaysia. *Asian Social Science*, *9*(5), 1–17.

Rajaratnam, S. D. (2018). Qualitative research in tourism: Reflections of an Asian researcher. In P. Mura, & C. Khoo-Lattimore (Eds.), *Perspectives on Asian tourism*, https://doi.org/10.1007/978-981-10-7491-2_13.

Reimer, J. K. (Kila), & Walter, P. (2013). How do you know it when you see it? Community-based ecotourism in the Cardamom Mountains of southwestern Cambodia. *Tourism Management*, *34*, 122–132.

Rowat, D., & Engelhardt, U. (2007). Seychelles: A case study of community involvement in the development of whale shark ecotourism and its socio-economic impact. *Fisheries Research*, *84*(1), 109–113.

Scheyvens, R. (1999). Ecotourism and the empowerment of local communities. *Tourism Management*, *20*, 245–249.

Scheyvens, R. (2000). Promoting women's empowerment through involvement in ecotourism: Experiences from the third world. *Journal of Sustainable Tourism*, *8*(3), 232–249.

Scheyvens, R. (2002). *Tourism for development: Empowering communities*. Prentice Hall.

Sebele, L. S. (2010). Community-based tourism ventures, benefits and challenges: Khama Rhino Sanctuary trust, central District, Botswana. *Tourism Management*, *31*, 136–146.

Sekaran, U. (2003). *Research methods for business: A skill building approach* (4th ed.). John Wiley and Sons, inc.

Sood, J., Lynch, P., & Anastasiadou, C. (2017). Community non-participation in homestays in Kullu, Himachal Pradesh, India. *Tourism Management*, *60*, 332–347. https://doi.org/10.1016/j.tourman.2016.12.007

Stone, L. S., & Stone, T. M. (2011). Community-based tourism enterprises: Challenges and prospects for community participation; Khama rhino sanctuary trust, Botswana. *Journal of Sustainable Tourism*, *19*(1), 97–114.

Stronza, A., & Gordillo, J. (2008). Community views of ecotourism. *Annals of Tourism Research*, *35*(2), 448–468.

Stronza, A., & Pêgas, F. (2008). Ecotourism and conservation: Two cases from Brazil and Peru. *Human Dimensions of Wildlife*, *13*, 263–279.

Su, M. M., Wall, G., Wang, Y., & Jin, M. (2019). Livelihood sustainability in a rural tourism destination-Hetu Town, Anhui Province, China. *Tourism Management*, *71*, 272–281. https://doi.org/10.1016/j.tourman.2018.10.019

Tao, T., & Wall, G. (2009). Tourism as a sustainable livelihood strategy. *Tourism Management*, *30*(1), 90–98.

Telfer, D., & Sharpley, R. (2008). *Tourism and development in the developing world*. Routledge.

The Tourism Sector and Sustainable Development Goals Report. (2017). Retrieved March 1, 2020, from https://unstats.un.org/sdgs/files/report/2017/thesustainabledevelopmentgoalsreport2017.pdf.

Tran, L., & Walter, P. (2014). Ecotourism, gender and development in northern Vietnam. *Annals of Tourism Research*, *44*, 116–130. https://doi.org/10.1016/j.annals.2013.09.005

Vaz, J., & Pyne, J. (1997). *The Kinabatangan Floodplain: An introduction*. WWF Malaysia.

Walter, P. (2011). Gender Analysis in Community-based Ecotourism, *Tourism Recreation Research*, *36*(2), 159–168, DOI:10.1080/02508281.2011.11081316.

Weinberg, A., Bellows, S., & Ekster, D. (2002). Sustaining ecotourism: Insights and implications from two successful case studies. *Society and Natural Resources*, *15*, 371–380.

Wu, T. C., Wall, G., & Tsou, L. Y. (2014). Serious tourists: A proposition for sustainable indigenous tourism. *Current Issues in Tourism*, *20*(13), 1–20. https://doi.org/10.1080/13683500.2014.970143

Zapata, M. J., Hall, C. M., Lindo, P., & Vanderschaeghe, M. (2011). Can community-based tourism contribute to development and poverty alleviation? Lessons from Nicaragua. *Current Issues in Tourism*, *14*(8), 725–749. https://doi.org/10.1080/13683500.2011.559200

Community-based ecotourism and the transformative learning of homestay hosts in Cambodia

Vicheth Sen and Pierre Walter

ABSTRACT
Ecotourism seldom places humans within the natural environments which serve as ecotourism attractions. Scholarship on learning and education in ecotourism has accordingly focused mostly on visitor learning about wildlife, biodiversity and adventure experiences in 'pristine' natural ecosystems. By contrast, community-based ecotourism (CBE) involves local and Indigenous communities living within natural environments. CBE aims not only to preserve natural environments, but also to conserve the traditional cultures of the peoples living within them. CBE involves visitors in longer term village homestays, local guides, and environmental, livelihood and cultural learning. It has also recently become an important area of study. However, while learning by CBE homestay hosts and guides is equally important, it is mostly unstudied to date. This paper reports on an empirical, interpretive case study of the transformative learning of CBE hosts in a rainforest-based CBE project in Chambok, Cambodia. Findings show that CBE hosts experienced transformations in their worldviews and self-identities in relation to three areas of transformative learning: environmental conservation, gender roles and local culture. These are three areas of learning beneficial to both hosts and communities, and key to the success of CBE. As such, they might well be encouraged in other CBE projects.

Introduction

Community-based Ecotourism (CBE) has proven itself a more sustainable, holistic and localised alternative to conventional forms of short-term ecotourism. Wildlife ecotourism (whale-watching, swim with the dolphins, turtle nesting, etc.) can be low impact and result in substantive conservation education for visitors, and benefits for wildlife, but for the most part it does not involve humans or local communities as stakeholders (Ballantyne, Hughes, Lee, Packer, & Sneddon, 2018; de Lima & Green, 2017). Likewise, nature ecotourism (nature walks, birding, interpretive walks) and adventure tourism (rock climbing, marine kayaking, mountaineering, surfing, scuba, whitewater rafting) also encourage an understanding and ethic of natural conservation, but again do not generally include the humans who live in or near pristine natural areas, including Indigenous communities (Mackenzie, 2017; Rantala, Rokenes, & Valkonen, 2018; Walter, 2013). In other cases, 'ecotourism' has simply been adopted as a marketing term for the 'greenwashing' of mega beach resorts, hotel chains, cruise ship tourism and mass ecotours detrimental to wildlife, local ecosystems and human communities (Das & Chatterjee, 2015; Honey, 2008; Self, Self, & Bell-Haynes, 2010). In contrast to other forms of ecotourism, CBE aims not only to conserve the natural environment and educate visitors, but also to provide cultural and economic benefits for the local people who live within these environments, thus initiating a positive cycle of mutually reinforcing conservation and community development (Masud, Aldakhil, Nassani, & Azam, 2017; Reimer & Walter, 2013).

CBE is an environmentally-friendly, human-implicated and relational form of ecotourism rich in opportunities for learning and education, by both visitors and hosts (Walter, 2009, 2013; Walter & Reimer, 2012). Typically, visitors travel to local and Indigenous villages in remote areas of great natural biodiversity and beauty – tropical rainforests, high mountain terrains, marine island communities – not only to view megafauna, megaflora and breathtaking scenery, but also to interact with the peoples who live in these areas. Visitors stay overnight in villages in the homes of local host families ('homestays') often for several days, or stay in tents or huts as they trek, hike or boat with guides who are also family homestay and community members. Visitors learn about the local environment, flora and fauna, natural foods and medicines, traditional livelihood practices, and the local and Indigenous cultures of their hosts. Unlike more conventional forms of ecotourism, in CBE, there is little separation among nature, livelihood and culture, and much of the authenticity of the experience comes in the learning relationships developed between visitors and hosts, who are also resident experts in all three areas.

For the most part, extant scholarship on learning and education in CBE has focused on learning by ecotourism visitors, as opposed to hosts, reflecting the focus of wider ecotourism research (Ballantyne et al., 2018; Walter, 2013). However, some literature has begun to identify the varied forms of learning performed by CBE hosts not only in learning how to develop and manage CBE homestay and ecotourism guiding operations, but also in the knowledge hosts gain about environmental conservation, livelihood practices and the revival of their own traditional cultures as they learn and teach these topics to CBE visitors (Regmi & Walter, 2016; Walter, 2009; Walter, Regmi, & Khanal, 2018). Most CBE learning is in fact a two-way street: visitors learn from hosts, but hosts also learn from visitors. CBE visitors are usually international tourists or domestic tourists from urban areas with different cultural norms, knowledge, beliefs and skills than hosts. CBE hosts are generally rural, village-based people who farm and fish, harvest forest and marine products, and often still largely adhere to traditional livelihood and cultural practices, although almost all are also integrated into modern cash, information technology and consumer economies as well. In short, host-visitor interactions in CBE comprise a tourism 'contact zone' (Pratt, 1991), rich in cross-cultural learning exchanges for both hosts and tourists (Dines, 2018; Minca, 2009; Somerville & Perkins, 2003).

To date, research on CBE in Cambodia has not focused on either visitor or host learning (one exception is Walter & Reimer, 2012). A recent review of 51 studies of sustainable tourism in Cambodia 1993–2013 found no studies directly focused on learning or education, but many in which host learning was identified as important to the success of CBE projects (Carter, Thok, O'Rourke, & Pearce, 2015). Research on CBE in Cambodia has included, for example, findings on a lack of community awareness of CBE benefits (Clements et al., 2008), inequitable participation and distribution of benefits along gender and other lines (Reimer & Walter, 2013), conflict with exploitive industries such as logging and fishing (De Lopez, 2003; King, Basiuk, Chan, & Yem, 2009), and the contribution of CBE to livelihood and forest conservation (Lonn, Mizoue, Ota, Kajisa, & Yoshida, 2018, 2019). While some of these concerns lie beyond the power of education to address, it is also true that CBE training programmes, on-the-job learning and community learning have (a) had good success in building community and homestay host and guide capacity to reduce environmental impacts; (b) allowed more equitable sharing of benefits; and (c) empowered communities to face outside threats such as logging, mining and mass tourism development (Reimer & Walter, 2013; Tran & Walter, 2014; Walter, 2009; Walter et al., 2018; Walter & Reimer, 2012).

In this paper, we are particularly interested in understanding the informal, transformative learning of homestay hosts in CBE. By 'informal learning,' we mean all of the adult learning which takes place outside of formal schooling; in the community, family, home and workplace (Marsick & Watkins, 2001; Schugurensky, 2000). In CBE, we take informal host learning as mostly experiential (learning by doing, trial and error, experimentation) and self-directed: there is no 'standard curriculum' for informal CBE host learning, and outside of training programmes, no formalised instructors, nor structured pedagogy with specified learning outcomes (Regmi & Walter, 2016; Walter et al., 2018). Further, as is the case with other adults, we understand that informal learning by hosts may be purposeful, and directed towards particular aims, but may also be unintended, incidental or gained through tacit processes of socialisation (Schugurensky, 2000). As such, host learning will vary in form and content across particular individuals, groups of learners, and the social and cultural contexts they inhabit.

By 'transformative learning,' we mean informal host learning which is catalyzed by a particularly meaningful experience or set of experiences to produce a change in adults' perspectives on the world, identities, social roles and actions (Mezirow, 2000, 2009). That is, we are interested in host learning which results in changes in the epistemologies which 'constitute and direct individuals' capacities and embodied knowledge, their sense of self (i.e. subjectivity) and gaze (i.e. how they view the world and how they believe the world is viewing them)' (Billett, 2014, p. 688). We are less interested in the 'curriculum practices' and 'pedagogical practices' of workplace learning (Billett, 2014) by CBE hosts; that is, the ways in which hosts learn the basic occupational practices of CBE in the workplaces of their homes, communities and surrounding environments (Regmi & Walter, 2016). These practical 'nuts and bolts' of CBE development, design and management have been fairly well-documented over the last two decades (Calanog, Reyes, & Eugenio, 2012; Denman, 2001; Fennell, 2014) and are not the focus of this paper. Instead we focus on the new knowledge, beliefs, worldviews and identities which CBE hosts gain through their participation in CBE projects; that is, hosts' transformative learning. In the next section, we explain transformative learning theory in CBE and review related research findings on transformative learning in tourism. This is followed by a description of our CBE study site of Chambok

Village in Cambodia, and our research methodology. We then present findings in relation to learning experiences of Chambok CBE hosts, followed by a discussion of findings in light of the literature. We conclude with implications for CBE research and practice.

Conceptualising transformative learning in CBE

As noted above, transformative learning involves a change in an adult's perspective on the world, and in many ways an epistemological shift in the way adults see themselves following a particularly meaningful experience or set of experiences. In Mezirow's (2000, 2009) model, adults are seen to move through a 10-stage process of transformative learning starting with a 'disorienting dilemma' which sparks an emotional self-examination, followed by a critical assessment of personal assumptions, recognition of shared discontent, and an exploration of possible new roles, relationships, and actions (Figure 1). They then plan a course of action, gain new skills and knowledge to implement the plan, try out new roles, build competence, and lastly reintegrate learning and new identities into their lives (Mezirow, 2009). In Mezirow's words, transformative learning refers to

> a process by which we transform our taken-for-granted frames of reference (meaning perspectives, habits of mind, mind-sets) to make them more inclusive, discriminating, emotionally capable of change, and reflective so that they may generate beliefs and opinions that will prove more true or justified to guide action. (Mezirow, 2000, pp. 7–8)

Transformative learning can also be characterised as an *epistemological practice* of workplace learning wherein new knowledge or worldviews are constructed based on observation, experience and reflection (Billett, 2014; Regmi & Walter, 2016).

The relationship between experiential learning and transformative learning proposed by Walter (2016a) for visitors to CBE can also be adapted to explain host learning. The left side of Figure 1 shows a representation of Kolb's (1984) Cycle of Experiential Learning. In this cycle, hosts have some sort of Concrete Experience in hosting CBE visitors (e.g. having a stranger with foreign customs staying in their homes, observing the environmental conservation or cultural practices of their guests, trying to communicate across language, gender, class and national differences). They then engage in Reflective Observation on this experience, may coming up with generalisations about it (Abstract Conceptualisation), and test out these generalisations through Active Experimentation and further experiences. For CBE hosts, as opposed to visitors, perhaps the most emotionally-charged Concrete Experiences occur in the realm of Culture (top concentric oval, upper left of Figure 1) and cross-cultural experiences; and less so in the realms of Adventure and Nature, as is the case with CBE visitors. When a CBE host has an emotionally-charged Concrete Experience, this may act as a catalyst to start the 10-stage transformative learning process (right column, Figure 1). This catalytic experience is termed as 'disorienting dilemma' by Mezirow (2000), one which disturbs an adult's previous ideas and frames of reference, either in dramatic fashion (divorce, death, job loss, etc.) or more incrementally as adults have smaller emotionally-charged experiences, and ponder, reflect and change more slowly over time (Cranton, 2016). In both cases, however, the learning is transformative, with Mezirow's (2000, 2009) latter stages of acquiring new skills and knowledge, trying on new roles, building competence and confidence, and reintegration of new perspectives into one's life the more visible 'outcomes' of this transformative process.

Research on transformative learning in tourism is still fairly sparse. However, several studies exist which have clear implications for CBE host learning. In study of transformative learning in volunteer tourism, Coghlan and Gooch (2011) found that volunteers overseas went through many of Mezirow's 10 stages, with disorienting dilemmas resulting from 'experiencing an unfamiliar cultural/social/natural environment' (p. 718). A more recent study by Coghlan and Weiler (2018) on the same topic established that volunteers' transformative learning is largely an internal, individualised process rather than a set of definable, or generalisable behavioural outcomes. An earlier study by Zahra (2011) reported that volunteers' encounters with the caste system, religion, and poverty were life-changing experiences which resulted in volunteers becoming less self-centered and materialistic, more aware of the importance of family and community, and more active in social justice work. Transformative learning for CBE visitors may be found in experiencing spiritual epiphanies in nature, physical and mental challenges of outdoor adventures, exposure to conditions of poverty, deforestation or mining, or the relational experience of *communitas* created in living together with local families in their village homes (Walter, 2016a, 2016b). Similar cross-cultural, nature and adventure experiences may also promote transformative learning by community hosts as they host, guide, teach and learn from CBE visitors in the contact zones of CBE.

As Christie and Mason (2003) note in discussing transformative learning for ecotourism guides, 'where the

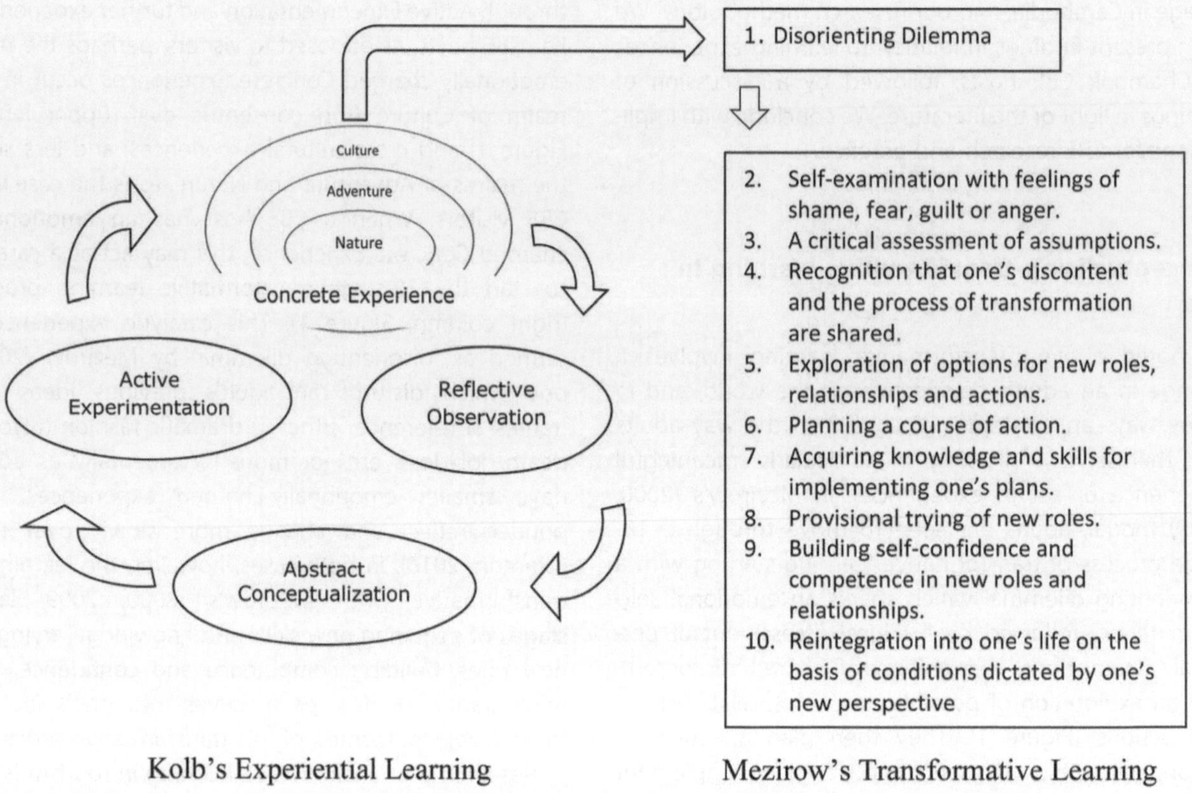

Figure 1. Experiential and transformative learning in CBE. Source: Walter (2016a, p. 1362).

cultural differences between visitors and members of local communities are particularly marked, then the transformation (on at least one side) is more likely to occur' (p. 1). In the present study of the transformative learning of Cambodian CBE hosts, we posit, again, that learning is in fact always at least two-way: visitors learn from hosts, but hosts also learn from visitors. Moreover, for hosts, learning is often not only individual, but collective learning at the family and community levels. That is, although CBE visitors may travel with friends and family members, in the ecotourism experience, visitors exist primarily as individuals outside of their own homes, families, workplaces and communities. By contrast, CBE homestay hosts fundamentally live, work, learn and educate visitors within their collective units and identities as they host CBE guests. Thus families and communities also learn and change with the development of CBE projects and the hosting of outside guests in CBE projects, even after visitors leave.

Research site

The Chambok CBE site is considered one of the most successful cases of CBE in Cambodia (Lonn, 2013). Chambok, popular among both urban Khmer and international visitors, is considered a 'model' ecotourism site, and its CBE project and village homestay accommodations have received numerous awards. These include the ASEAN Homestay Standard Award in 2016 and finalist in the 'Engaging Communities' category of the National Geographic World Legacy Awards in 2017 (Chambok, 2019; Duch, 2016). The Chambok CBE project has been extensively studied (e.g. Lonn, 2013; Lonn et al., 2018, 2019; Men, 2006; Va, Lay, & Chhum, 2008; Ven, 2015; Walter & Reimer, 2012) and has its own CBE website and social media. Chambok is located about 110 kilometres west of Phnom Penh via national road No. 4, in a rather isolated mountainous region next to the protected area of Kirirom National Park, Kampong Speu province (Figure 2). Kirirom National Park has been a prime recreational site for urban Cambodian elites since the 1950s (Rith, Williams, & Neth, 2009), except for the civil war period. Based on the spatial division of ecotourism regions in Cambodia (Walter & Sen, 2018), Chambok belongs to the *mangrove and rainforest* region of southwestern Cambodia. The Chambok ecotourism catchment area is located in Chambok commune, and comprises nine villages in total, with 520 families and 1,828 people, virtually all of whom engage in rice farming as their primary occupation (Chambok, 2019; Ven, 2015).

During the genocidal reign of the Khmer Rouge (1975–1979), the population of Chambok Commune was forcibly relocated to the market town of Treng Trayoeung, and the area of Chambok became a

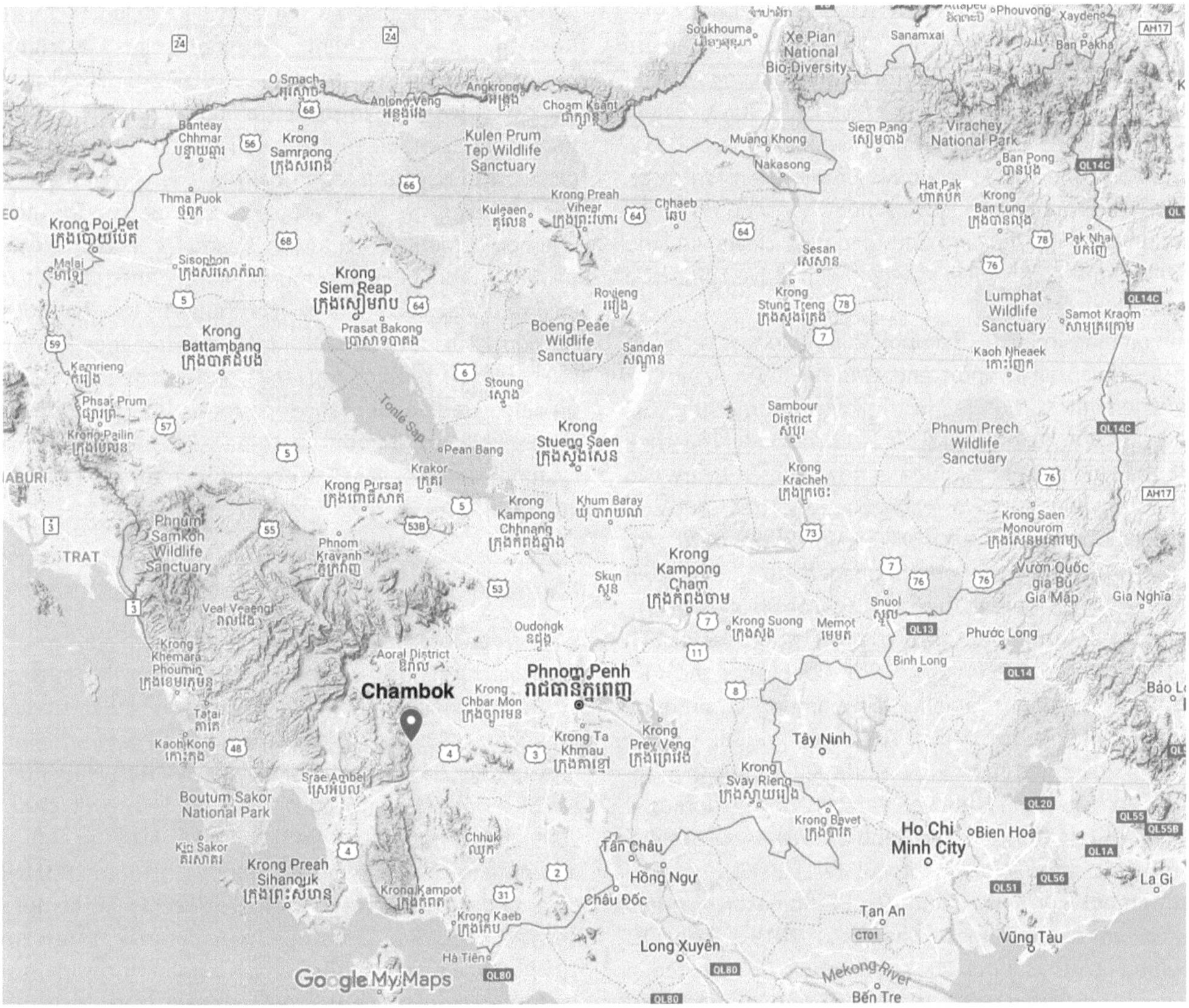

Figure 2. Research site.
Source: The diagram sourced from Walter 2016a is based on Kolb (1984), Mezirow (2000).

stronghold for the Khmer Rouge (Lonn et al., 2018). Khmer villagers were only able to return to Chambok in 1998, following the end of factional fighting in the country. The present Chambok CBE project was established four years later, in 2002 (Chambok, 2019). Today, about half of Chambok villagers have joined the CBE project, and have received CBE and homestay training, and initial CBE infrastructure improvement support, including trail and road construction, homestay renovations and land mine clearance (Chambok, 2019; Va et al., 2008).

The initial impetus for the establishment of the CBE project in 2002 was a desire to address the environmental degradation of Chambok area forests and lands contiguous to and inside of Kirirom National Park. These rainforest lands and ecosystems had experienced extensive illegal logging, hunting of wildlife, and harvest of wood for charcoal and fuelwood production, carried out by both local villagers and outside investors (Rith et al., 2009). Land mines left from the Khmer Rouge military and civil war also meant that many areas were unsafe for villagers and ecotourism. The overarching aim of the Chambok CBE project was to provide alternative sources of income to logging, hunting and wood extraction, to build local capacity for sustainable forest management, and to raise awareness of local villagers and visitors about the importance of environmental conservation (Lonn, 2013; Ven, 2015). Initiated by a local Cambodian environmental non-governmental organisation called *Mlup Baitong* (Green Shade), the Chambok project officially opened to visitors in 2003, and gradually expanded over the next 5–6 years (Va et al., 2008). During this time, CBE development activities included construction of ecotourism infrastructure, capacity building and training of the CBE management committee members and homestay hosts, training for ecotourism food, guiding and hosting services and management, measures to promote environmentally-friendly ecotourism, and support for women's self-help and savings groups (Steck, 2013). All project costs were funded by various foreign donors including Oxfam Novib, Scottish Catholic International Aid Fund, Oxfam Great Britain, Asian Development Bank, McKnight Foundation and the UNDP/Economic Commission (Va et al., 2008). The Cambodian government contributed some support for road building and landmine clearance, and private tour companies assisted with advertising (Va et al., 2008).

Beginning in 2010, Chambok CBE no longer received outside financial support and was run by a local CBE management committee with advice from the NGO Mlup Baitong (Lonn, 2013). The 15-member Chambok CBE management committee is now elected every five years and consists of a leader, deputy, treasurer and representatives from each village in Chambok (Lonn, 2013). Nine of the 15 management committee members are women. The main responsibilities of the management committee is to 'oversee ecotourism activities, maintain the ecotourism sites, look after the CBT fund and distribute benefits fairly among community members' (Lonn, 2013, p. 3). The Chambok CBE project today is managed mostly by ethnically Khmer Cambodians and, as noted above, attracts mostly domestic Khmer visitors from Phnom Penh as well as substantial numbers of international visitors (Steck, 2013).

Chambok currently offers 42 CBE homestays spread across all nine villages (Chambok, 2019). From its opening in 2003 until 2013, Chambok CBE hosted a total of 136,050 visitors, both local and international (Steck, 2013). In 2013 alone, 10,000 people visited the Chambok CBE project, 25 percent of whom were international visitors from about 30 different countries, who generated approximately USD 41,000, representing a tenfold increase in annual revenues since the project started (Steck, 2013). Chambok currently has an average of about 1,000 visitors per month, of which about 200 are international tourists (Mlup Baitong, 2019). In 2019, 40% of community CBE revenues went for conservation, 25% to supporting vulnerable member in times of need, 20% for a community development fund and 15% for a reserve fund (Impact Explorer, 2019). The community development fund supports women's self-help groups, health and micro-enterprise initiatives, and the community water supply system (Mlup Baitong, 2019). Chambok CBE also hosts numerous study tours for other Cambodian villages and organisations interested in developing CBE (about 50 per year), further testament to Chambok's success as a model CBE site (Walter & Reimer, 2012).

Ecotourism attractions in the Chambok CBE project include a 40-meter high waterfall and three smaller waterfalls, and other nature activities including forest trekking, ox-cart riding, cycling, bird watching, camping, swimming at the base of a waterfall, visiting bat caves, learning to make traditional crafts, watching traditional Cambodian dances, learning to cook local dishes, and participating in daily village life. Homestays are, however, the main source of CBE income for most Chambok families. Figures 3 and 4 show typical Chambok homestay accommodations. Other sources of income that ecotourism contributes to the local economy include entrance fees, parking fees, ox-cart rides, bicycle rentals, souvenirs and handcrafts, tourist guiding, traditional dance performances by primary school children, and an all-women restaurant run by a rotating roster of local women (Chambok, 2019; Men, 2006).

Methodology

We took an interpretive case study approach appropriate to researching host learning in CBE projects; that is, to an area of scholarship which has been little studied (Merriam, 1998; Yin, 2017). The aim was to gain a preliminary understanding of the dimensions of host learning in a CBE homestay project, framed by Mezirow's (2000, 2009) basic notions of transformative learning. Again, appropriate to a new area of study, this interpretative case study approach was not designed to be comprehensive, representative or generalisable (Merriam, 1998); the aim was rather to have a point of comparison for further studies in other CBE contexts. As such, recruitment of participants was purposeful, not random or representative. Study participants were selected as key informants for their intense involvement and extensive experience in the CBE project at large, and in homestay hosting in particular. By our reckoning, they were thus more likely to have experienced transformative learning during the CBE project, and more likely to be able to reflect on this learning in light of their knowledge of the CBE project. Study participants meeting these criteria were recommended to us by members of the CBE management committee. These participants included the head of the CBE site, the head cook, the head of the women's savings group, the homestay coordinator, a local CBE guide, and four other long-term homestay accommodation hosts without additional leadership roles in the project (Table 1).

We used basic qualitative research methods employed in case study research to collect our data. These included an exhaustive review of previous research on the Chambok CBE project and project documents, and semi-structured interviews with key informants in the CBE project. Interviews were conducted at

Figure 3. A typical homestay at Chambok.

the Chambok CBE site in December 2016. Data collection included participant observation in the nine homestay villages in the Chambok CBE commune, and an overnight stay with a host family in their home. Interviews were conducted in Khmer, transcribed, translated into English, and then coded for themes related to transformative learning.

In recruiting study participants, we were not concerned if their perspectives were 'biased' or non-representative (Merriam, 1998): we were looking for personal meaning within local context, for their individual stories of transformation, for insights into how they saw the experience of learning in CBE hosting; we expected that these experiences, perspectives and

Figure 4. Sleeping amenities at a Chambok homestay.

Table 1. Research participants.

Identifier	Gender	Role
C1	Man	Homestay coordinator
C2	Woman	Head cook and president of women's savings group
C3	Man	Head of the CBET site
C4	Woman	Homestay host
C5	Woman	Homestay host
C6	Woman	Homestay host
C7	Woman	Homestay host
C8	Woman	Guide

personal reflections would differ from person to person. Thus, notions of validity and reliability of findings were not particularly germane in the interview data collected from participants: we trusted that interviewees conveyed the personal truths that they wanted to convey about their CBE learning experiences. Where validity and reliability did come into play, however, was in the ways in which we as researchers analyzed and interpreted the data presented to us. As such, in data analysis, we used a replicable 'constant-comparative process' (Merriam, 1998) in which the composite set of all interview data was first reviewed to identify tentative themes related to host transformative learning in relation to the study purpose. Our guiding questions were: What new knowledge, beliefs, worldviews and identities did CBE hosts gain through their participation in the Chambok CBE project? Where did they experience transformative learning, with what results? Themes generated in response to these questions were then checked against the composite data set and refined into three mutually exclusive study findings presented in the next section. Data from participant conservation, previous research and CBE documents were used mainly to provide local understanding and context to the findings generated in interviews. Finally, it is important to note that an elaboration of Mezirow's 10 stages of transformative learning by CBE hosts was beyond the scope of the present paper. Instead, we present findings on the latter stage outcomes of transformative learning for hosts' experience in Chambok CBE hosting.

Findings

Three areas of transformative learning were identified for hosts in the Chambok CBE project: (1) environmental conservation, (2) changes in gender roles and relations, and (3) rediscovering local culture. The first finding represented the fulfilment of an early objective in the design of the CBE project in the early 2000s, as noted above. By providing alternative income through CBE activities, homestays and ancillary businesses, and environmental education to hosts, the aim was not only to move villagers away from extractive forest industries such as logging and hunting, but also to initiate forest protection and regeneration measures through community forest patrols, conservation fees, a tree nursery and tree planting. These Chambok conservation aims are a basic premise of most CBE projects (Reimer & Walter, 2013). Further, it was hoped that an ethic and community awareness of the benefits of environmental conservation, sanitation and hygiene would be gained through knowledge learned and revived in teaching and learning from Mlup Baitong and ecotourists, from the implementation of new systems of recycling and solid waste disposal, and from the provision of clean water, toilets and sewage systems, again serving ecotourists, but also benefitting the Chambok community. The second finding of transformative learning through changes in gender roles and relations was likewise promoted by Mlup Baitong in Chambok project CBE training to bring greater self-confidence, income, community involvement and leadership roles to women, as has been the case in other successful CBE projects (Tran & Walter, 2014; Walter, 2011). The third finding of transformative learning and cultural revival, a key component of other CBE projects (Regmi & Walter, 2016; Reimer & Walter, 2013), was promoted in the Chambok CBE design as well.

Environmental conservation

The experience of hosting outside visitors in homestays, guiding them on ecotours, and teaching them local knowledge of rainforest, wildlife, landscape, riverine ecosystems and livelihood systems profoundly changed the attitudes and behaviour of Chambok hosts in relation to environmental conservation. It transformed their views about their relations to the surrounding environments and helped to revive, and value, traditional ecological knowledge and forest conservation practices. One homestay host, typical of others, explained how the CBE project and the livelihood opportunities provided by homestays had allowed them to shift their thinking away from income earning practices destructive of forest to a more sustainable livelihood dependent on healthy forests:

> Before I would go logging or collecting forest by-products and sell them to support my living. Now that I run a homestay, I stay home and take care of my grandchildren. I can earn a living from hosting visitors and contribute to the ecotourism community .... I think the environment is better conserved now because we do not go logging and destroy it. It's been nearly ten years since I last went logging and collecting forest products. I used to destroy forests by logging and collecting other forest products. That was my way of earning a living at the time. That was the past. But now I

> understand the importance of the forest and the environment. (C6)

A second host noted that the CBE project had also protected the Chambok forest from outsider logging and land grabs:

> I think the ecotourism community has helped raise general awareness about the importance of the forest. This has in turn helped reduce illegal logging. This ecotourism community also protects the community forest from being encroached upon or taken by certain powerful and wealthy people. (C2)

The first host quoted above (C6), typical again of other hosts, further commented on her transformative learning in relation to environmental sanitation and cleanliness as a result of participating in the CBE project and hosting visitors, and change within the community:

> Before becoming involved in hosting visitors, we did not care much about anything. We did not understand anything about the work we're doing now. We did not care about how good or bad the environment was. We did not care about sanitation or cleanliness. We simply did not care and threw things all over the place. We did not have toilets and defecated all over the place. We were not aware of sanitation or anything like that. After becoming part of the ecotourism community and starting to host visitors, there has been a lot of change. I now have my own toilet. I keep my house clean. It's not 100% perfect yet, but there has been change in my thinking and living. By doing so, I am living a healthy life, not contracting any disease ... I now understand the importance of sanitation. It's important for my children and grandchildren, myself, and visitors. (C6)

In hosting both domestic and foreign visitors, some hosts also observed that their guests had different environmental behaviours than they did, and reflected on this as part of the learning of environmental conservation; in this case in relation to littering. One of the hosts interviewed noted that: 'foreign guests don't throw away trash all over the place. Khmer guests are different. Some keep their trash properly. Others don't. The hosts take care of that' (C7). Other homestay hosts felt very proud of their clean and sanitary homes and villages, and were admired by both visitors and fellow villagers, as well as being named 'model' homestay hosts. One of the hosts spoke about how she had changed her view of her home after starting to host visitors, saying:

> (I went through) a lot of change. Before I began hosting visitors, I didn't know how to arrange my house, keep my house clean, cut down bushes around my house, or collect plastic bags and water bottles around my house. But now my house is very clean. I have kept my house clean, and I am happy because when people see my house, they admire it as a well organized house which is kept clean. (C4)

Two of the hosts interviewed had in fact recently received Association of Southeast Asian Nations (ASEAN) Homestay Awards (meeting 75–100% of criteria) honouring homestays for standards of management, safety, hygiene and social and cultural sustainability (ASEAN Secretariat, 2016, p. 25), reinforcing their sense of pride at how they had changed their homes. The Chambok CBE visitor centre also served as a learning resource for both visitors and hosts, with a small exhibit of Chambok, its flora and fauna, a history of the development of CBE in the community.

Changes in gender roles and relations

In many respects, while CBE homestay hosting provides new sources of income for women, men and families, it also involves an intensification of women's traditional domestic labour in cooking, cleaning, and hosting guests, on top of their normal workloads in caring for their families (Tran & Walter, 2014; Walter et al., 2018). However, at the same time, even with increased domestic labour, Chambok women homestay hosts found new self-confidence and pride in running homestay accommodations. One woman homestay host, for example, commented on her involvement in the CBE programme and hosting: 'My husband had frequently attended training and meetings at the Community. I have rarely attended anything. I am usually at home, cleaning and preparing things around the home' (C7). When asked who did most of the work at home when there were guests coming to stay at her house, she said it was her. Yet she also took great pride in her new capability to run a homestay business:

> I don't want to sound boastful, but I think I am quite good [at hosting visitors]. ... I realize I have some value. Previously, I thought I was useless because I am illiterate. But now, since the establishment of the ecotourism community, I have received this ASEAN Homestay Award. I am very happy. (C7)

A second woman host put it this way:

> Before, I thought I did not have any value or importance. I considered myself of low value. But now, after I have been able to do what I have done (running a homestay), I consider myself of high value. I didn't know of my value before. I thought I was not capable of doing anything on my own. But now that I have been able to do what I have done, I realize I have value. (C6)

She went on to say that, although she was largely playing the domestic role in hosting visitors, 'I have gained confidence in myself, knowing that I can do it on my own. I'm not 100% capable yet, but I have confidence in my capacity' (C6). Another Chambok homestay host elaborated on her learning experience:

> Compared to the time before I joined the (CBE project), I have learned a lot about community development. Previously I didn't know anything about community work; after harvest season ended, I would go to make additional income by logging and collecting forest by-products. But now after harvesting, I do community work. I think that now I can contribute to community development and can still generate additional income for my family. (C2)

Overall, the hosting of visitors in both homes and the community was a transformative learning experience for many of the women engaged in this work. It not only expanded their traditional domestic gender roles, but also provided them new skills, knowledge and opportunities for leadership and community involvement. The head of the women's kitchen and women's savings groups, for example, received extensive kitchen management and culinary training from the NGO Mlup Baitong, including field visits to other communities. She learned how to schedule shifts, plan and prepare simple financial reports, manage the kitchen and the savings groups and cook different dishes. Now, whenever she travels or encounters new dishes, she tries to learn how to prepare them and then cooks them in the ecotourism kitchen when she returns to Chambok. When she does not understand something, she asks her peers or staff from Mlup Baitong when they visit. She reflected on the transformative change in self-identity which she experienced in her new community roles:

> Since I started participating in the CBE project, even though I am not literate, I have been sent on study tours to two provinces. Although I am not literate, I can remember what we do in our community. That is why I was sent to a study tour in Kampot province. Although I am not literate, I have realized that I have my own value and I am worthy of recognition by others. I have given myself value. It's completely different from before I joined the CBE. I have grown a little bit. Before, I was not only illiterate but also ignorant about everything. But now I understand my value. Although I am not literate, I am valued by other people. I can do something, like making souvenirs, cooking different dishes, and hosting visitors, and the visitors also value me. I have learned about and given myself value, which I am very happy about. I realized I have value for society like others. (C2)

Revaluing local culture

Hosting visitors as part of the ecotourism project in Chambok also provided very important learning for hosts in relation to reviving local culture and traditions at the community. Prior to the CBE project, for example, traditional Khmer dances were not valued the same way as they are now. The fact that the Khmer Rouge Pol Pot regime killed 90% of all Khmer classical dance performers and other artists from 1975 to 1979, as members of a 'decadent class' (Falser, 2014; Grant, 2014), meant that the revival of traditional dance in Chambok by children dancers was particularly poignant and meaningful for the community. The Chambok homestay coordinator, for example, commented that 'the ecotourism project makes possible the traditional dances, which helps the community to maintain our traditional … culture' (C1). Since the dances themselves embodied aspects of traditional Khmer village beliefs and practices (harvest, courting rituals and so on), this was also important cultural content being revived. Moreover, Chambok dancers were often asked to teach visitors some of the Khmer dances, thus reinforcing these traditional dance cultures for both guests and hosts.

The head of Chambok CBE played a central role in raising awareness among all villagers in the community of the importance of maintaining their local culture and traditions. He constantly reminded the young dancers, for example, of the importance of the traditional dances performed to welcome visitors and the rationale behind them, and to not perform them only for the sake of earning money. The children (six boys and six girls) clearly valued their dancing and performed the dances with pride, especially when asked to teach guests. Many of the Chambok villagers interviewed spoke about how their involvement in hosting visitors as part of the ecotourism community not only allowed them to generate income but also contributed to maintaining their local traditions and practices. These traditional practices included the care of oxen and the use of ox carts to give visitors rides, the learning and teaching of traditional ecological and farming knowledge (names and uses of plants, how to cultivate and harvest rice, fruits and vegetables), and the making and teaching of traditional rattan jewellery (bracelets, rings and necklaces).

Learning to cook various forms of Khmer food, as well as teaching cooking classes to visitors using local ingredients over open fire stoves was also an important form of cultural revival and education provided by the CBE project. Hosts might teach visitors how to make a Khmer stir-fried dinner meal, or traditional bamboo stuffed with rice, mung bean and coconut roasted over the fire, or might take their guests on a tour of household noodle-making, gardening and rice wine production. The head CBE cook talked about learning how to prepare and appreciate various Khmer dishes, mostly local, at CBE training sessions arranged for cooks, and then sharing her knowledge and skills with other cooks at the women's kitchen at Chambok. As she recounted,

> I learned to cook from trainers sponsored by Mlup Baitong and through learning from other cooks. I also

> learned from a cook, who was a personal chef for a group of Singaporean visitors. Besides, I know how to cook local dishes based on my own experience of preparing food for my family at home. The MSME (Micro, Small and Medium Enterprises) project also provided me with an opportunity to observe other cooks preparing different dishes in Phnom Penh. It was a two-week training and observation. (C2)

In terms of other cultural learning, the homestay coordinator observed that international visitors often asked him for suggestions about proper behaviour so that they would not upset local people or offend their traditional and cultural norms. One young woman CBE guide observed that there were areas of mutual cultural learning, for example, in environmental conservation, and appropriate clothing to wear when swimming in a pool downstream from the waterfall:.

> Based on my experience guiding foreign visitors, first, most of them love the environment. Most of them are very knowledgeable about the environment. But, secondly, their culture in terms of clothing is very different from our Khmer culture in rural, mountainous areas like Chambok. Completely different. It is normal for them to go swimming in bikinis; sometimes they don't even wear anything when they go swimming. As a guide, I don't wear a swimsuit like they do. I usually wear sports outfit or shorts, but not a bikini. They once asked me "Why don't you have a bikini to wear? Do you normally go swimming in this clothing?" I replied, "Yes. In rural areas we normally wear like this to go swimming.... we don't normally wear bikinis or short clothes. In fact, my shorts are already too short. We respect our culture. Based on our traditions, we normally wear long clothes." That's one cultural difference. (C8)

In addition to reviving traditional cultural traditions and lifeways, the experience of managing the Chambok CBE project and their homestays has taught the CBE leaders and hosts to adapt to changes in their communities, and to find new ways to attract visitors. The head of the CBE project talked about future plans to expand the site to keep the flow of visitors coming to the community and to showcase and preserve the local culture and traditions:

> Our future plans? We plan to build four homestays in the forest to offer natural lodging, which has been requested by many visitors. We also plan to build a local library to display local agricultural tools and other local materials, traditions and cultural practices to maintain our local culture. (C3)

Discussion and conclusion

Findings of this study illustrate how learning in CBE can create transformations in host worldviews and self-identities in the three areas of environmental conservation, changes in gender roles and relations, and the revival of local culture. These are areas of learning both beneficial to hosts and their communities and important to the success of CBE projects. The first set of findings on environmental conservation were similar to findings of other research on CBE host learning in Cambodia and Nepal (Reimer & Walter, 2013; Walter et al., 2018). These findings indicated the success of the Chambok CBE project in meeting its major goal of forest conservation, regeneration and protection. Homestay host learning about solid waste, sewerage, sanitation and clean accommodations was also important to hosts and instilled a sense of pride and accomplishment in their homestay accommodations. Second, in learning to take on new gender roles and responsibilities in homestay hosting and the CBE project at large, Chambok women villagers also experienced transformative learning: they felt empowered, and realized their own value not only in effectively managing homestays and generating substantial income for their families, but also in new leadership roles they adopted in the Chambok community. Again, this demonstrated the effectiveness of the Chambok CBE project design. Findings on gender were similar to other CBE research in diverse community settings in Cambodia, Nepal, Peru, Vietnam, Indonesia, Guyana and the Philippines (Dilly, 2003; Schellhorn, 2010; Stronza, 2005; Tran & Walter, 2014; Walter et al., 2018). Finally, the Chambok project allowed CBE homestay hosts to revive cultural practices and traditions, like Khmer dance, traditional cooking, and Buddhist and forest spirit rituals largely forbidden to them during the Khmer Rouge's reign of terror some 40 years ago. They felt pride in their ability to learn, exchange and teach cultural practices and a sense of *communitas* with visitors, similar to other CBE hosts in Nepal, Thailand, Cambodia and Australia (Higgins-Desbiolles, 2009; Walter et al., 2018; Walter & Reimer, 2012).

The study had limitations in that it focused on the outcomes of transformative learning experienced by Chambok CBE hosts and was therefore both limited in scope and depth; it did not explain the processes or stages of transformative learning. We do not know, for example, what discrete disorienting dilemmas, if any, sparked individual host's processes of transformative learning along the lines of Mezirow's (2000, 2009) theory. We can locate in the findings above, research participant quotes which perhaps suggest emotionally-charged experiences (disorienting dilemmas). These include a woman Chambok guide first seeing naked and bikini clad foreigners swimming, and a non-literate woman CBE cook constantly referring to her self-identity with the words 'although I am not literate,' followed by 'I am not ignorant; I, the community, the project and visitors have recognized my value; I have been on study tours' and so on. What we can identify here, again without knowing exactly what provoked these changes for this woman, are likely some of the further stages of Mezirow's transformative learning theory: perhaps an emotional self-examination, a critical assessment of personal assumptions, an exploration of possible new roles, relationships and actions, the planning of a course of action to gain new skills and knowledge. Based on her interview, this CBE host had clearly tried out new roles, built competence and reintegrated learning and new identities into her life (Mezirow, 2009) as a CBE Cook, host and teacher of others, and overcome any obstacles presented by non-literacy in learning to do so. We might perhaps see here an epistemological shift whereby this capable woman constructed new knowledge, a new worldview and a new self-identity based on observation, experience and reflection (Billett, 2014; Regmi & Walter, 2016). Likewise, with further questioning, in the CBE host participant quote above of 'I used to destroy forests by logging and collecting other forest products ... now I understand the importance of the forest and the environment,' we might discover the source of this change, and the process of how this participant came to change his beliefs, livelihood practices and behaviour towards environmental conservation.

This study has merely established the fact that transformative learning occurred in several key areas for some Chambok hosts purposefully selected for the likelihood of this learning happening. It does not tell us about the experiences of the other 36 Chambok homestay hosts and their learning experiences, transformative or otherwise, nor how transformative learning may have occurred for long-term hosts like the eight study participants over the last 17 years of the Chambok CBE project. These are both limitations of the study and valuable areas of further research. Knowing that study participants did experience what appear to be fairly profound life and identity changes as a result of their CBE hosting is important in theoretical terms, as it confirms the theoretical endpoint of Mezirow's stages of transformative learning (2000, 2009), and suggests that there are likely further stages to be identified through more research. The study has also identified three areas of host transformative learning which support CBE project aims, thus suggesting support (albeit in limited fashion) for the Chambok CBE project aims, design and management.

The findings on environmental conservation, gender roles and relations, and cultural revival may well serve as a point of comparison and departure for several types of future research. First, it would be worthwhile to conduct follow-up interviews with the eight participants in the study to ask them about the stages and processes of transformative learning. Limitations of time and resources did not allow this in the present study. Walter's (2016a) model of experiential and transformative learning for visitors in CBE might serve as a useful guide to this end, tying CBE hosting experience to transformative learning. Second, the eight participants might be expanded to include other CBE hosts who indicate they have experienced transformative learning, including those who are further recommended by current study participants. An initial household survey of all 42 CBE host families might likewise be conducted to identify both those hosts who indicate strong pathways or stages of transformative learning, and those who do not. A sample of each group could then be recruited for in-depth interviews as to how transformative learning was or was not promoted in their CBE hosting experience based on key family, community, project and other contextual barriers and supports.

In sum, although limited in scope, this study serves as an important first step in bringing the voices of local CBE hosts – as opposed to visitors and other outsiders – to bear on questions of adult learning in CBE. In conceptual terms, it identifies three transformative learning outcomes and possible (speculative) processes for host learning in CBE. In cautious terms, these study findings speak to the efficacy of CBE projects like Chambok which have adopted extensive and varied forms of nonformal education and training – workshops, study tours, demonstrations, etc – and have allowed numerous opportunities for hosts to pursue their own self-directed informal learning in the areas of environmental conservation, gender empowerment and cultural preservation. Based on study findings, these are CBE educational designs and policies which should likely be continued and/or implemented in other projects. In broader terms of CBE as community development, the poignant testimony of at least one of the non-literate woman CBE hosts quoted above, and that of the other study participants also possessing little formal education, speaks to the capacity, courage and persistence of local and Indigenous adults involved in CBE projects. While such adults may possess limited schooling, they possess vast knowledge of natural ecosystems, traditional livelihood, culture and local relationships in their roles as CBE hosts (Reimer & Walter, 2013; Schellhorn, 2010; Walter, 2009). How best to understand, honour and value this local knowledge and epistemologies in CBE projects, visitor and host learning, including transformative learning, is thus also a critical question for future study.

Disclosure statement

No potential conflict of interest was reported by the authors.

Funding

This work was supported by Social Science and Humanities Research Council of Canada: [Grant Number 15R20398].

ORCID

Pierre Walter http://orcid.org/0000-0001-7168-5260

References

ASEAN Secretariat. (2016). *ASEAN homestay standard*. Jakarta, Indomesia: ASEAN Secretariat.

Ballantyne, R., Hughes, K., Lee, J., Packer, J., & Sneddon, J. (2018). Visitors' values and environmental learning outcomes at wildlife attractions: Implications for interpretive practice. *Tourism Management*, *64*, 190–201.

Billett, S. (2014). Learning in the circumstances of practice. *International Journal of Lifelong Education*, *33*(5), 674–693.

Calanog, L. A., Reyes, D. P. T., & Eugenio, V. F. (2012). *Making ecotourism work. A manual on establishing community-based ecotourism enterprise (CBEE) in the Philippines*. Makati City: Philippines Office, Japan International Cooperation Agency.

Carter, R. W. B., Thok, S., O'Rourke, V., & Pearce, T. (2015). Sustainable tourism and its use as a development strategy in Cambodia: A systematic literature review. *Journal of Sustainable Tourism*, *23*(5), 797–818.

Chambok. (2019). *Chambok ecotourism site*. Retrieved from https://chambok.org/home/

Christie, M. F., & Mason, P. A. (2003). Transformative tour guiding: Training tour guides to be critically reflective practitioners. *Journal of Ecotourism*, *2*(1), 1–16.

Clements, T., John, A., Nielsen, K., Chea, V., Ear, S., & Meas, P. (2008). *Case study: Tmatboey community-based ecotourism project, Cambodia*. USAID Translinks. Retrieved from https://rmportal.net/conservation-enterprises/ce-documents/tmatboey-community-based-ecotourism-project-cambodia/view

Coghlan, A., & Gooch, M. (2011). Applying a transformative learning framework to volunteer tourism. *Journal of Sustainable Tourism*, *19*(6), 713–728.

Coghlan, A., & Weiler, B. (2018). Examining transformative processes in volunteer tourism. *Current Issues in Tourism*, *21*(5), 567–582.

Cranton, P. (2016). *Understanding and promoting transformative learning: A guide to theory and practice* (3rd ed.). Virginia: Stylus.

Das, M., & Chatterjee, B. (2015). Ecotourism: A panacea or a predicament? *Tourism Management Perspectives*, *14*, 3–16.

de Lima, I. B., & Green, R. J. (Eds.). (2017). *Wildlife tourism, environmental learning and ethical encounters: Ecological and conservation aspects*. Cham, Switzerland: Springer.

De Lopez, T. T. (2003). Economics and stakeholders of Ream National Park, Cambodia. *Ecological Economics*, *46*(2), 269–282.

Denman, R. (2001). *Guidelines for community-based ecotourism development*. Gland, Switzerland: WWF International.

Dilly, B. J. (2003). Gender, culture, and ecotourism: Development policies and practices in the Guyanese rain forest. *Women's Studies Quarterly*, *31*(3/4), 58–75.

Dines, N. (2018). An irreconcilable first-place: The precarious life of tourism and heritage in a southern European historic centre. *International Journal of Heritage Studies*, *24*(2), 142–153.

Duch, L. (2016, January 6). A success story: Chambok Community-based Ecotourism. *Khmer Times*. Retrieved from https://www.khmertimeskh.com/34244/a-success-story-chambok-community-based-ecotourism/

Falser, M. S. (2014). From a colonial reinvention to postcolonial heritage and a global commodity: Performing and re-enacting Angkor Wat and the Royal Khmer Ballet. *International Journal of Heritage Studies*, *20*(7-8), 702–723.

Fennell, D. A. (2014). *Ecotourism*. London, UK: Routledge.

Grant, C. (2014). Perspectives of culture-bearers on the vitality, viability and value of traditional Khmer music genres in contemporary Cambodia. *The Asia Pacific Journal of Anthropology*, *15*(1), 26–46.

Higgins-Desbiolles, F. (2009). Indigenous ecotourism's role in transforming ecological consciousness. *Journal of Ecotourism*, *8*(2), 144–160.

Honey, M. (2008). *Ecotourism and sustainable development: Who owns paradise?* (2nd ed.). Washington, DC: Island Press.

Impact Explorer. (2019). *Chambok ecotourism community*. Retrieved from https://impactexplorer.asia/project/chambok-community-project/

King, P., Basiuk, R., Chan, B. S., & Yem, D. (2009). *Strengthening sustainable tourism: Strategic environmental assessment of the tourism sector in Cambodia*. Bangkok: Asian Development Bank.

Kolb, D. A. (1984). *Experiential learning*. Englewood Cliffs, NJ: Prentice-Hall.

Lonn, P. (2013). Can community-based ecotourism improve local livelihoods? A case study of Chambok community-based ecotourism, Southwest Cambodia. *Cambodia Development Review*, *17*(3), 1–7.

Lonn, P., Mizoue, N., Ota, T., Kajisa, T., & Yoshida, S. (2018). Evaluating the contribution of community-based ecotourism (CBET) to household income and livelihood changes: A case study of the chambok CBET program in Cambodia. *Ecological Economics*, *151*, 62–69.

Lonn, P., Mizoue, N., Ota, T., Kajisa, T., & Yoshida, S. (2019). Using forest cover maps and local people's perceptions to evaluate the effectiveness of community-based ecotourism for forest conservation in Chambok (Cambodia). *Environmental Conservation*, *46*(2), 111–117.

Mackenzie, S. H. (2017). Adventure tourism. *Tourism Review International*, *21*(3), 221–329.

Marsick, V. J., & Watkins, K. E. (2001). Informal and incidental learning. *New Directions for Adult and Continuing Education*, *89*, 25–34.

Masud, M. M., Aldakhil, A. M., Nassani, A. A., & Azam, M. N. (2017). Community-based ecotourism management for sustainable development of marine protected areas in Malaysia. *Ocean & Coastal Management, 136*, 104–112.

Men, P. (2006). Tourism, poverty, and income distribution: Chambok community-based ecotourism development, Kirirom National Park, Kampong Speu province, Cambodia. *Journal of GMS Development Studies, 3*(1), 25–40.

Merriam, S. B. (1998). *Qualitative research and case study applications in education*. San Francisco, CA: Jossey-Bass.

Mezirow, J. (2000). Learning to think like an adult. In J. Mezirow & Associates (Eds.), *Learning as transformation: Critical perspectives on a theory in progress* (pp. 3–33). San Francisco, CA: Jossey-Bass.

Mezirow, J. (2009). Transformative learning theory. In J. Mezirow, E. W. Taylor, & Associates (Eds.), *Transformative learning in practice: Insights from community, workplace, and higher education* (pp. 18–31). San Francisco, CA: Jossey-Bass.

Minca, C. (2009). The island: Work, tourism and the biopolitical. *Tourist Studies, 9*(2), 88–108.

Mlup Baitong. (2019). *Visit our ecotourism sites: Chambok*. Retrieved from http://mlup-baitong.org/ecotourism_more/

Pratt, M. L. (1991). Arts of the contact zone. *Profession, 91*, 33–40.

Rantala, O., Rokenes, A., & Valkonen, J. (2018). Is adventure tourism a coherent concept? A review of research approaches on adventure tourism. *Annals of Leisure Research, 21*(5), 539–552.

Regmi, K. D., & Walter, P. (2016). Conceptualising host learning in community-based ecotourism homestays. *Journal of Ecotourism, 15*(1), 51–63.

Reimer, J. K., & Walter, P. (2013). How do you know it when you see it? Community-based ecotourism in the Cardamom Mountains of southwestern Cambodia. *Tourism Management, 34*, 122–132.

Rith, S. O., Williams, P. W., & Neth, B. (2009). Community-based ecotourism and rural livelihood diversification: Reframing the approach. In P. Beaupre, J. Taylor, T. Carson, H. Keam, & C. Heng (Eds.), *Emerging trends, challenges and innovations: Community based natural resource management (CBNRM) in Cambodia* (Vol. II, pp. 471–492). Phnom Penh, Cambodia: Learning Institute.

Schellhorn, M. (2010). Development for whom? Social justice and the business of ecotourism. *Journal of Sustainable Tourism, 18*(1), 115–135.

Schugurensky, D. (2000). The forms of informal learning: Towards a conceptualization of the field. *NALL Working Paper No. 19*. Toronto, Canada: Ontario Institute for Studies in Education, University of Toronto.

Self, R. M., Self, D. R., & Bell-Haynes, J. (2010). Marketing tourism in the Galapagos Islands: Ecotourism or greenwashing. *International Business & Economics Research Journal, 9*(6), 111–125.

Somerville, M., & Perkins, T. (2003). Border work in the contact zone: Thinking Indigenous/ non-Indigenous collaboration spatially. *Journal of Intercultural Studies, 24*(3), 253–266.

Steck, B. (2013). Award winner: Chambok community-based ecotourism project. Retrieved from www.todo-contest.org/preistraeger-en/pdf/kambodscha-2014-e.pdf

Stronza, A. (2005). Hosts and hosts: The anthropology of community-based ecotourism in the Peruvian Amazon. *Napa Bulletin, 23*(1), 170–190.

Tran, L. T., & Walter, P. (2014). Ecotourism, gender and development in northern Vietnam. *Annals of Tourism Research, 44*, 116–130.

Va, M., Lay, K., & Chhum, S. (2008). Good practice in the Chambok community-based ecotourism project in Cambodia. In P. Steele, N. Fernando, & M. Weddikkara (Eds.), *Poverty reduction that works: Experience of scaling up development success* (pp. 3–19). London, UK: Earthscan.

Ven, S. (2015). Host residents' attitude toward community-based ecotourism: Empirical study in Southwestern Cambodia. *Journal of Tourism & Hospitality, 4*(1), 1–11.

Walter, P. (2009). Local knowledge and adult learning in environmental adult education: Community-based ecotourism in southern Thailand. *International Journal of Lifelong Education, 28*(4), 513–532.

Walter, P. (2011). Gender analysis in community-based ecotourism. *Tourism Recreation Research, 36*(2), 159–168.

Walter, P. (2013). Theorising visitor learning in ecotourism. *Journal of Ecotourism, 12*(1), 15–32.

Walter, P. (2016a). Catalysts for transformative learning in community-based ecotourism. *Current Issues in Tourism, 19*(13), 1356–1371.

Walter, P. (2016b). Travelers' experiences of authenticity in "hill tribe" tourism in northern Thailand. *Tourist Studies, 16*(2), 213–230.

Walter, P., Regmi, K. D., & Khanal, P. R. (2018). Host learning in community-based ecotourism in Nepal: The case of Sirubari and Ghalegaun homestays. *Tourism Management Perspectives, 26*, 49–58.

Walter, P., & Reimer, J. K. K. (2012). The "ecotourism curriculum" and visitor learning in community-based ecotourism: Case studies from Thailand and Cambodia. *Asia Pacific Journal of Tourism Research, 17*(5), 551–561.

Walter, P., & Sen, V. (2018). A geography of ecotourism in Cambodia: Regions, patterns, and potentials. *Asia Pacific Journal of Tourism Research, 23*(3), 297–311.

Yin, R. K. (2017). *Case study research and applications: Design and methods* (6th ed.). Thousand Oaks, CA: Sage.

Zahra, A. (2011). Volunteer tourism as a life-changing experience. In A. M. Benson (Ed.), *Volunteer tourism: Theoretical frameworks and practical application* (pp. 90–101). Oxon, UK: Routledge.

Gender analysis in community-based ecotourism

Pierre Walter

Abstract: Community-based Ecotourism (CBET) holds great promise in promoting environmental conservation, local livelihood, and cultural preservation. However, without a clear understanding of the differential benefits and costs of CBET for men and women, 'gender unaware' CBET research, policies and projects may in fact promote the welfare of men over women. That is, without a consideration of factors such as (a) the gender division of labour, (b) gender relations, and (c) differential access to and control over environmental, livelihood and cultural resources, CBET projects may unwittingly exacerbate existing gender inequalities in local communities. As a result, CBET projects may fail to meet their fundamental environmental, livelihood and cultural aims. Following a brief discussion of CBET, this paper provides a rationale for gender analysis in ecotourism, outlines the historical evolution of gender and development work, and introduces standard frameworks and tools of gender analysis. Using extant empirical case studies of CBET, the paper then discusses how both 'efficiency' (Women in Development) and 'empowerment' (Gender and Development) tools of gender analysis might be applied in the research and practice of Community-based Ecotourism.

Introduction

With the exception of work by Scheyvens (2000, 2007) and Swain and Swain (2004), there has been little theoretical engagement with gender issues in the field of ecotourism. The general gender 'blindness' of mainstream ecotourism theorizing is evident, for example, in a recent comprehensive review of research in the field (Weaver and Lawton 2007). In appraising some 75,000 abstracts in 6,000 periodicals pertinent to ecotourism, Weaver and Lawton (2007) extract only two studies of ecotourism related to gender: a study by Fennell (2002) identifying gender differences in motivation and behaviour among Canadian ecotourists in Costa Rica (2007: 1172), and a study by Belsky (1999) which found 'discrepancies in class, gender and patronage' in community-based ecotourism in Belize (2007: 1174).

Likewise, with the exception of Higham (2007a), standard texts and edited collections in ecotourism, and those related to Community-based Ecotourism (CBET) in particular—such as Honey (2008), Stronza and Durham (2008) and Zeppel (2006)—make scant mention of gender issues. In short, in the field of ecotourism, there appears to be little awareness of the existence and importance of gender perspectives to understand key dimensions of environmental conservation, community livelihood and cultural preservation promoted through Community-based Ecotourism.

Standard concepts and tools of gender analysis have for several decades been productively employed in a broad range of disciplines related to Community-based Ecotourism. These include, but are not limited to, gender and tourism (Aitchison 2005; Tucker 2007; Nakata and Momsen 2010; Fergusen 2010), gender analysis in natural resources management (Kelkar *et al.* 2003; Vernooy 2006; Currie and Vernooy 2010), gender analysis in the prevention and management of invasive species (Fish *et al.* 2010), gender analysis of plant breeding in natural resource management (Lilja and Ashby 1999), and analysis of the gendered impacts of climate change (Kelkar 2009). By comparison, theorizing and analysis of gender issues in ecotourism is largely behind the times: the field appears to be somewhat of a (male) dinosaur amidst the larger world of gender-aware environmental conservation and sustainable development.

The purpose of this paper is to consider how several standard frameworks and tools of gender analysis might be applied to research in Community-based Ecotourism. While research on other forms of ecotourism—wildlife viewing, adventure tourism and so on—might also benefit from a gender analysis, the frameworks presented here are most appropriate for CBET projects in development contexts. In the following sections, Community-based Ecotourism is first defined in relation to 'authentic' ecotourism (Honey 2008), community-based tourism (Suansri 2004) and indigenous ecotourism (Zeppel 2006). Second, a rationale is given for why a gender analysis should be included in research on CBET. Third, a brief overview of the historical evolution of gender analysis in development work—moving from Women in Development (WID) to Gender and Development (GAD) to Social and Gender Analysis (SAGA)—is presented. Examples of established gender analysis frameworks are then illustrated using recent CBET case studies which include gender as an analytical category. The paper concludes with an appeal to adopt gender analysis tools in the research and practice of Community-based Ecotourism.

Community-based Ecotourism

The meaning of ecotourism continues to be widely debated (Weaver 2005; Donohue and Needham 2006; Cater 2006; Higham 2007b). In their review of ecotourism research, Weaver and Lawton (2007: 1170), however, argue that even while diverse definitions of ecotourism abound, these share three core criteria: '(1) attractions should be predominantly nature-based; (2) visitor interactions with those attractions should be focused on learning or education, and (3) experience and product management should follow principles and practices associated with ecological, socio-cultural and economic sustainability'. Community-based approaches to ecotourism, in distinguishing themselves from the wider category of ecotourism, are initiatives which share these three criteria, 'but which also aim to ensure that members of local communities have a high degree of control over the activities taking place, and a significant proportion of the benefits accrue to them' (Scheyvens 1999: 246). In defining standards of practice in ecotourism certification, Honey and Stewart (2002: 1-2) argue for a similar operational definition:

> ...ecotourism is a multifaceted concept that involves travel to fragile, pristine, and usually protected areas. It strives to be low impact and (usually) small scale; helps educate the traveller; provides funds for conservation; directly benefits the economic development and political empowerment of local communities; and fosters respect for different cultures and human rights.

In her seminal work, *Ecotourism and Sustainable Development*, Martha Honey (2008) calls this form of ecotourism 'authentic' ecotourism to distinguish it from less community-based forms of ecotourism (i.e., wildlife, adventure and nature). Other authors name this 'community-based tourism', allowing for environmental conservation, but stressing economic livelihood and cultural preservation (Suansri 2004). Community-based approaches to ecotourism also have many shared premises with 'indigenous ecotourism' (Zeppel 2006). Common to all three notions of ecotourism are: (a) principles of local participation, control or ownership of ecotourism initiatives; (b) a focus on environmental conservation and local livelihood benefits; (c) the promotion of customary and indigenous cultures; and to some extent, (d) the promotion of local and indigenous human rights and sovereignty over traditional territories and resources (Table 1). For the purposes of this paper, this common core will be characterized as 'Community-based Ecotourism' or CBET.

Table 1. Variations of Community-based Ecotourism

'Authentic' Ecotourism Honey (2008)	**Community-based Tourism** Suansri (2004)	**Indigenous Ecotourism** Zeppel (2006)
▪ Involves travel to natural destinations ▪ Minimizes environmental and cultural impact ▪ Builds environmental awareness ▪ Provides direct financial benefits for conservation ▪ Provides financial benefits and empowerment for local people ▪ Respects local culture ▪ Supports human rights and democratic movements	▪ Recognizes, supports and promotes community ownership of tourism ▪ Involves community members from the start in every aspect ▪ Promotes community pride ▪ Improves the quality of life ▪ Ensures environmental sustainability ▪ Preserves the unique character and culture of the local area ▪ Fosters cross-cultural learning ▪ Respects cultural differences and human dignity ▪ Distributes benefits fairly among community members ▪ Contributes a fixed percentage of income to community projects	▪ Based on indigenous knowledge systems and values ▪ Promotes indigenous customary practices and livelihoods ▪ Used to retain rights to access, manage and use traditional lands and resources ▪ Used to manage cultural property such as historic and sacred sites ▪ Operates under the control and active participation of local indigenous people ▪ Involves indigenous communities in ecotourism planning, development and operation ▪ Terms of trade are negotiated for the use of ecotourism resources, including people

Why a Gender Perspective?

An engagement with gender issues and gender analysis in tourism studies at large has been on-going since at least the mid-1990s (Kinnaird and Hall 1994, 1996; Swain 1995). Early on, in their introduction to *Tourism: A Gender Analysis*, Kinnaird *et al.* (1994: 24) elaborated a 'gender-aware framework' with three basic concepts: '(1) tourism development processes and tourism-related activities are constructed out of gendered societies; (2) gender relations both inform, and are informed by, the practices of all societies; and (3) power relations surrounding tourism development processes represent an extension of the politics of gender relations'. The collection of case studies which followed in the book then showed how an analysis of gender roles and relations in tourism development, depending on social context, can in some cases benefit women, but in others, disadvantage them.

On the one hand, in Kinnaird and Hall's (1994) edited collection, there are gender analysis case studies in contexts where social relations in tourism development act simply to reinforce common norms of patriarchy found in larger 'gendered societies' (in this case, England, Ireland and the Caribbean). In these instances, women's participation in low wage, low skilled tourism labour functions as an intensification of traditional, 'invisible' domestic labour for women tourism hosts — in cleaning, cooking and serving for tourists: this labour is 'long hours of what amounts to a lot of drudgery, doing work which men refuse to do...' (Breathnach *et al.* 1994: 70). A rise in various forms of work in the sex trade also occurs across both genders. On the other hand, in other research (in this case, rural Spain), a gender analysis showed that a 'commoditization' of women's domestic work in the hosting of tourists was largely beneficial to women, and in fact allowed them a measure of economic and personal gain, and ultimately, the survival of family farms (Swain 1995: 260). Other case studies of tourism development in societies with traditional norms of greater gender equity (such as Western Samoa) demonstrated how women are able to 'capitalize on the opportunities the (tourism) industry offers because their rights have been safeguarded by customary norms' (Fairbairn-Dunlop 1994: 139). In short, these and other early studies demonstrate how a 'gender-aware framework' helps to reveal the importance of understanding gender roles and relations in tourism development.

Since its inception in the 1990s, theoretical work on gender and tourism studies has provided an increasingly complex and varied analysis of changes in gender identities, roles and relationships in tourism development (e.g., Kinnaird and Hall 2000; Hall *et al.* 2003; Aitchison 2005; Pritchard *et al.* 2007; Bianchi 2009; Fergusen 2010; Aitchison in press). In a far-ranging review of theory on gender and feminist perspectives in tourism studies, Aitchison (2005) describes the recent 'cultural turn' of post-structuralist theorizing. This cultural turn examines issues such as hegemonic discourses of tourism, the construction of (oppressive) gender identities, inequalities of power within cultural practices, tourist and host subjectivities, embodiment, sexuality and symbolic representation in tourism. Aitchison (forthcoming) notes a synergy emerging between this more recent post-structural analysis and longer standing materialist, structural approaches. In this hybrid tradition, a case study of gender and tourism development in Turkey, for example, shows how (Muslim) women, who have traditionally been proscribed from participating in the public sphere, have taken on new gender roles in paid tourism work outside of the private sphere of the home, and as a result, have 'crafted' new gender identities and inspired new gender relations (Tucker 2007). Likewise, a case study by Cabezas (2006: 507) shows how tourism labour in Cuba's beach resorts is 'sexualized'; that is, for workers to receive tourism benefits, they must 'enact scripts of affective labour, sexual enticement, and servility in the delivery of their occupational roles'. Moreover, these scripts are stratified along lines of race, color, sexuality and gender.

In recent empirical research on ecotourism, structural traditions of gender analysis are predominant, although the cultural turn in Tourism Studies is also acknowledged (Dilly 2003; He 2003; Walter 2004; Kelkar 2004; Stronza 2005; Pleno 2006; Di Ciommo 2007; Dowley 2007; Flacke-Neudorfer 2007; Gentry 2007; León 2007; Weinert 2008; Schellhorn 2010). In one of the few conceptual framings of gender and community-based ecotourism, Scheyvens (2000, 2007) proceeds from a participatory development lens to argue, on the one hand, that in many ecotourism contexts, women have greater interaction with the natural environment than men, and thus possess specialist environmental knowledge, and also different interests in the conservation of particular environmental resources from men. Moreover, women are often at the forefront of efforts to preserve local culture. On the other hand, women often have lower status, poorer paid employment in ecotourism, or simply engage in unpaid labour as an extension of their domestic work. At the community level, Scheyvens (2007: 193) notes that gender-unaware 'projects intended to benefit an entire community have inadvertently disadvantaged women'; when, for example, income is paid to (male) heads of household and does not reach women. Finally, as Scheyvens tells us, although ecotourism can in some contexts open up new possibilities for women to participate in household and community decision-making, in other contexts, this new political power is resisted by men and other vested interests.

Gender and Development

In the history of gender analysis in international development work, there has been an evolution moving from the Welfare Approach in the 1970s, to the Women in Development (WID) in the mid-1970s and mid-1980s, to

Gender and Development (GAD) in the late 1980s and mid-1990s, and finally to the present-day focus on Social and Gender Analysis (SAGA) (Leach 2003; Currie and Vernooy 2010). Briefly, the Welfare Approach aimed to aid women primarily in their traditional reproductive roles as wives and mothers through the provision of development initiatives such as nutrition education, training in home economics, family planning and so on. The WID approach then argued, however, that focusing exclusively on women's reproductive roles only perpetuated existing gender inequalities, and that women were in fact an underutilized resource for economic development (Razavi and Miller 1995). Women's critical work in agricultural production and trade was now recognized as vital to family livelihood. Development programmes were, therefore, initiated to provide inputs such as technology and credit to support women's productive labour, while at the same time encouraging small-scale income-generating activities for women (e.g., handicrafts, small businesses) and employment opportunities.

By the late 1980s, it was becoming evident that simply 'adding' women into the development mix (WID) was insufficient, and that efforts needed to be made to address the role of unequal gender relations in restricting women's access to development resources and in excluding them from the development process. It also became increasingly clear that WID's central focus on economics had caused a neglect of welfare concerns (i.e., women's reproductive roles). Moreover, there was a recognition that women's labour included not only productive and reproductive labour, but also community labour (organizing social, religious and ceremonial events, community improvement activities, etc.). Together these were known as women's 'triple role' (Moser 1993), or their 'triple burden', as others referred to them. At the same time, the utility of the family household as a common unit of analysis for production and consumption also questioned, and various models of heterogeneous households proposed (Sen 1990; Moser 1993). With a better understanding of women's social, economic and political subordination to men also came an analytical distinction between 'practical gender needs' such as food, water, housing, education, income and healthcare, and 'strategic gender needs' which arise as a result of women's subordination in society (Molyneux 1985; Moser 1993). The latter may include such needs as abolishing unfair divisions of labour, alleviating the burden of childcare, securing rights to land and property, obtaining access to credit, allowing women choice over child bearing, and taking measures to stop violence against women.

In recent iterations of gender analysis such as SAGA, the emphasis on gender has given way to a more holistic consideration of the intersections between gender and class, caste and ethnicity, social structures, hierarchical relations, material basis of culture, heterosexual gender norms, and above all, the importance of local and outside contexts in the success of projects (Vernooy 2007; Currie and Vernooy 2010). In the SAGA approach, no predetermined gender analysis tools or frameworks are applied; instead, social research, and capacity building for local 'gender experts' comes first. These local gender experts then determine what tools of analysis are appropriate for their particular communities.

Although there is a certain chronology in the unfolding of these approaches, each strand of gender analysis can also be found today in various forms and locations across the field of development studies and in the practices of development work. In the main, a distinction can be made between WID 'efficiency' approaches and GAD / SAGA 'empowerment' approaches to development (Razavi and Miller 1995; Leach 2003). Under WID, resources should be allocated to programmes and projects in the most efficient manner possible, to both men and women, depending mostly on their economic roles. That is to say, ignoring women as an untapped development resource is 'inefficient', while investing in them as human resources increases their economic productivity. GAD / SAGA as empowerment or 'participatory' approaches also encompass an analysis of labour, time and the allocation of resources across genders and can promote economic development. However, their principal aim is the transformation of gender relations to increase women's participation in, control over, and benefits from the development process. Greater 'agency' or power exercised by women in an action-oriented framework may subsequently result not only in increased income and other economic benefits, but also better access to education, health, political rights; reductions in labour burdens, and changes in oppressive cultural norms for women and other disadvantaged groups in society.

Gender Division of Labour

Basic tools used to analyze the gender division of labour, and women's participation in CBET projects are found in the Harvard Framework, a WID efficiency approach developed in collaboration with USAID in the 1980s (Moser 1993; Leach 2003). Among Harvard research tools are the Access and Control Profile, and the Activity Profile. Other related research tools include devices such as a daily, timed log of activities, a yearly calendar of labour, and mobility maps showing the spatial boundaries within which women lead their lives (March *et al.* 1999). The Harvard Access and Control Profile examines who has productive resources such as land, labour, technology, capital, income, credit and education. The Profile then distinguished between access to a resource and control over decisions about its allocation and use, and is valuable in determining how women are hindered from participating in each stage of a project cycle. The Activity Profile Matrix (Figure 1) captures the gender division of labour directly, in reproductive, productive and community roles, and is a very useful analytical device in researching and planning ecotourism ventures to include and benefit women. On completition, this matrix will provide information on who does what by gender, and in more detailed form, when, where and for how long they do this. Thus, it is possible to identify women's and men's distinct gender roles, and to further break these down into boys and girls, and other sub-categories based on ethnicity, age or class or by spatial location (home, community, region) if desired (Leach, 2003)

Roles	**Women / girls**	**Men / boys**
Productive		
Reproductive		
Community		

Source: Leach (2003): 37

Figure 1. Activity Profile Matrix

If ecotourism researchers and planners (in the efficiency tradition) expect women to participate in and benefit from projects, they might use these simple tools to systematically document what constraints and encouragements to participation women face (e.g., childcare, cooking, health, food production), how they use the natural environment (in agriculture, aquaculture, collection of water and firewood, gathering of forest, field and marine foods and products, etc.), and what outside resource inputs will change this equation. What labour and time commitments, resources, and mobility limitations prevent or encourage local women (and others) from participating in ecotourism projects? How can women—who in many societies take on the lioness's share of reproductive labour, for example, find 'extra' time, energy and labour to plan ecotourism projects, coordinate hosting, act as guides and so on? What labour-saving and time-saving devices or resources might be introduced here, and to what effect? How do and might men and boys, by taking on activities such as childcare, cleaning, cooking and so on, help free up women's labour for ecotourism? With this basic information, researchers, planners and funders of ecotourism projects can then determine what particular inputs and changes are needed to provide some assurance of project success for both men and women.

To help illustrate how a gender analysis under the Harvard Framework might be applied, a preliminary accounting of the gender division of labour — reproductive, productive and community — from research on Community-based Ecotourism is now presented.

Reproductive Labour: An analysis here encompasses not only how gender roles proscribe, limit or encourage participation in ecotourism projects, but also how reproductive labour contributes to the business of ecotourism. Recent empirical research shows that when women participate in CBET, as in tourism at large, the hospitality function of hosting ecotourists in homestays, ecolodges and restaurants often means an intensification of women's reproductive labour. That is, women's domestic tasks are commodified and extended as service for tourists (feeding guests, collecting fuelwood, hauling water for food and bathing, cleaning guest rooms, making beds, etc.). The intangible, background 'emotional work' of hosting, hospitality and dispute resolution (making guests feel comfortable, safe and welcome) is also important reproductive labour here (Brandth and Haugen 2007). In this sense, a 'housewifization' of employment for women may occur in ecotourism development (Gentry 2007). In a study of a village ecotourism initiative adjacent to Gunung Rinjani National Park in Lombok, Indonesia, for example, Schellhorn (2010) found that while women were employed in the local hospitality businesses (guesthouses, restaurants) in fair numbers, their status was mostly as low or no wage family 'helpers'. By contrast, 90% of formally employed workers in local hospitality businesses were men, who in fact had less training than women, but received better wages. In general, benefits from ecotourism employment flowed mostly to men and incoming migrants, passing over local Sasak women. Another study of both mass tourism and alternative tourism, including ecotourism in Belize found that 'hotel domestic workers, waitresses, and cooks were generally female, while the majority of tour guides, taxi drivers, boat operators, and maintenance workers were male' (Gentry 2007: 484).

In a further dimension extending women's reproductive roles, ecotourism development has also led to women (and others) being pulled into the sex trade, as is at times the case in tourism development at large. As Kelkar (2004) tells us, for example, the development of ecotourism in Mosuo communities around Lugu Lake in Yunnan, China, has resulted in a commodification of (reproductive) cultural practices and a growing trade in women's bodies for sex.

Productive Labour: In CBET, this encompasses labour supplied to the production of food, shelter, and culture, to ecotourism guiding services, and to the many business functions of ecotourism management. In food production, for example, it is important to realize that there are 'women's' and 'men's' crops, livestock and cropping systems, as well as a division of food production tasks by gender (Sachs 1999; Kusujiarti and Tickamyer 2000). Local, subsistence, vegetable and root crops may be women's; market grains, trees, and non-food crops may be men's. In the gathering of non-timber forest and marine products – including food, medicine, herbs, packing and handicraft materials – and of shellfish, seaweed and other food in marine environments, the main collectors are usually women (Kelkar and Nathan 2003; Scheyvens 2007). Women and men, therefore, have different specialist knowledge and incentives to conserve different types of natural resources, different labour requirements in producing food, housing and crafts for ecotourists, and different arenas in which to demonstrate or teach livelihood tasks and local culture (e.g., language, dance, weaving, religious rituals).

Since CBET activities for ecotourists often involve not only homestays, but also participation in livelihood activities and enjoyment of cultural practices, women's productive labour in these areas is often intensified, as is the case with their reproductive labour. The same may be true for men, but in different arenas of labour. In the central labour of ecotourism guiding, for example, men usually predominate, in effect intensifying their traditional roles in hunting, fishing, boating and transport. By contrast, because of cultural and safety restrictions imposed on women's mobility, women are seldom allowed the opportunity to become guides. Women who do lead ecotours may be labelled as prostitutes, be subject to the threat of male violence, and may simply be told they are not strong enough for vigorous guiding in the wilderness or on the water (Scheyvens 2007; Schellhorn 2010). In fact, women may face patriarchal barriers to working as guides in their communities even when they are preferred by female ecotourists (EWN 2011). By contrast, in the labour of food and handicraft vending to ecotourists, women often predominate, in some cases following traditional gender roles, but in others breaking new ground and providing a new source of income and mobility for women (Stronza 2005; Pleno 2006).

Community Labour: The work of organizing and running CBET cooperatives, attending meetings, deciding on policies and procedures, resolving disputes, communicating among members, distributing benefits, marketing, participating in training programmes, coordinating with tour companies, representing projects to outsiders, and lobbying for political support is often men's labour, reflecting a gendered public-private split in spheres of activity. In CBET development in Mosuo communities in Yunnan, for example, the public labour of marketing ecotourism, securing bank loans, and coordination with tour businesses is increasingly done by men, while women now have intensified reproductive roles in housework, cooking and reception of guests (He 2003). In another project in a Hani community in Xishuangbanna, China, women serve tea, explain local history and customs, and sell goods to visiting tour groups, while men negotiate tours and act as guides in a community-run nature preserve (Walter 2004). However, in cases where ecotourism planning and education involves both men and women from the start, women may be drawn more equitably into community labour where they did not venture before. In a CBET project in Bohor, Philippines, for example, both men and women participated in ecotourism planning activities such as gender sensitivity workshops, training for communications, human relations, food production and hosting skills, and environmental conservation seminars (Pleno 2006). The result was greater mobility, decision-making and participation in community labour for women in the project. In short, determining the gender division of labour in CBET is a complex calculation based heavily on local contexts, but crucial in researching and understanding the gendered impact of CBET projects.

Gender Relations and Empowerment

Under Gender and Development (GAD) approaches, an analysis of the gender division of labour might be extended to include an examination of gender relations in CBET, and areas of change in these relations. One prominent example of such GAD approaches is Longwe's Women's Empowerment Framework (March *et al.* 1999; Leach 2003). This framework incorporates positive, neutral, and negative assessments of five levels of increasing equality and empowerment for women, moving from women's welfare to access, conscientization, participation, and finally, to

Table 2. Levels of Empowerment and Gender Needs

Longwe Levels of Empowerment			Moser
Control			Strategic Gender Needs
Participation	⇧	⇧	
Conscientization	Increasing Equality	Increasing Empowerment	Practical Gender Needs
Access			
Welfare			

Control – equal control in decision-making over factors of production and distribution of benefits

Participation – equal participation in decision-making processes related to policy-making, planning and administration

Conscientization – conscious understanding of gender roles and a gender division of labour that can be changed, and is fair and agreeable

Access – equal access to the factors of production (land, labour, credit, education, public services)

Welfare – equal access to material welfare (food, income, shelter, medical care).

Adapted from Leach (2003): 57-59

women's control over the development process. As noted above, a distinction might also be made between women's practical gender needs and strategic gender needs (Moser 1993). In Longwe's framework, practical gender needs correspond to welfare and access empowerment levels, while strategic gender needs are equivalent to Longwe's conscientization, participation and control levels (Table 2).

In an analysis of gender relations and levels of empowerment in CBET projects, both the importance of local context and the complexity of change becomes clear. For example, in the Mosuo communities around Lugu Lake, Yunnan, mentioned before, ecotourism development led to new gender roles which produced more income, but less political power for women (He 2003; Kelkar 2004). This situation roughly corresponds in Longwe's empowerment framework to better Welfare and Access to resources for women, but less progress on the levels of Conscientization, Participation and Control (Table 2). As He (2003: 164) tells us, even while Mosuo women adopt new entrepreneurial roles in hosting, guiding and management of ecotourism, and men contribute to reproductive labour in collecting firewood, 'the power of men is being consolidated and strengthened'. Mosuo men's specialist knowledge in natural resource management is also being valued over women's, and men are thus able to take advantage of training opportunities which reinforce their decision-making power in the household and community.

Another study, in this case of a community-based ecolodge in the Peruvian Amazon, found that when men went to work in the ecotourism business, women were left behind to assume men's labour in addition to their own (Stronza 2005). This was an additional labour burden for women, but it also opened up new gender roles and decision-making power. The ecotourism project thus created an unequal gender division of labour, but gave women more control in decision-making and greater involvement in community affairs, nominally moving them up Longwe's levels of empowerment (Stronza 2005: 185):

> Women whose husbands and sons were involved (in the ecolodge project) were generally spending more time alone as they maintained the household. In addition to their normal chores, which included a full day of washing clothes in the river, cooking, caring for young children, and keeping chickens and other small livestock, several women also became responsible for attending meetings in their husbands' stead, working on the farm, and selling produce in Puerto Maldonado. For every man who became involved in tourism, there was a woman who was taking on more and more responsibilities in the household, essentially becoming a single head of household.

In further research on ecotourism projects in the two fishing and agricultural communities in Bohor, Philippines (noted before), Pleno (2006) found that women's participation in CBET increased their skills, knowledge, self-confidence, mobility, political participation and environmental awareness, again moving them up Longwe's levels of empowerment from simple access to resources to greater participation and control. In the project, women served as guides for a mangrove boardwalk and on river tours, cultivated oysters for consumption and sale, and offered livelihood demonstrations such as weaving and cooking. They also worked to keep natural areas clean, halted firewood cutting, planted mangroves, and lobbied against the killing of birds. For some, this comprised a change in gender roles and relations: their husbands took on housework and childcare while the women attended meetings or worked as ecotourism guides.

The employment of women as ecotourism guides has been widely seen as a means to create changes in gender roles and increase women's levels of empowerment, but with mixed results. Schellhorn (2010), for example, in her study of CBET development in a Sasak village in Lombok, Indonesia, reports that, until recently, ecotourism guides were almost exclusively men. However, with outside funding, a training programme for local women guides was initiated, together with the establishment of a series of ecotourism day-trips to sites such as a traditional village, gardens, rice fields and a waterfall. However, even while these shorter trips were designed especially for less mobile women guides, nothing prevented men guides from simply adopting the shorter walks into their ecotourism itineraries, thus short-circuiting the initiative. Partly for this reason, very few women guides remained in the project over the long term. A local educational initiative in Nepal, however, had much greater success in training women ecotourism guides. Entitled Empowering Women in Nepal (EWN), the award-winning training programme has produced over 400 women alumni, many of whom are now much sought after as ecotourism guides (Jackson 2010; EWN 2011). The programme's aim is both to create economic opportunities in ecotourism for women and to protect the natural environment. Women's livelihood as ecotour guides is promoted by providing them training in English, tourism, trekking information, first aid, food preparation, history, geography, culture, mountaineering skills and business management. They also receive leadership training, and learn to be 'physically and mentally tough, and to approach their life with confidence at work and in their home' (Jackson 2010: 5). Loan programmes and links to alumni networks are instituted to support women's ecotourism businesses as well. Environmental awareness is promoted through training in health, hygiene and sanitation, food hygiene, composting, anti-pollution campaigns, crop rotation, waste management and the use of alternative fuel sources. Finally, in a direct action aimed at overturning oppressive gender relations in Nepal, EWN has established a safehouse and school for female child labourers sold into indentured servitude, many of whom are identified and rescued by women ecotourism guides trained in the programme.

Conclusion

Almost three decades now into its existence, ecotourism has expanded its reach into numerous locales and areas of the world, into academic publications, reports, projects and educational programmes (Weaver and Lawton 2007). Research on community-based ecotourism, as a particular variant of sustainable tourism, has also grown in leaps and bounds (Honey 2008). However, in neither case has there been much of a sustained engagement with gender issues in theorizing or research, in spite of lively debates in the larger 'mother' fields of tourism studies and sustainable development. Community-based ecotourism is potentially a very powerful model of sustainable tourism in which the often contradictory aims of economic livelihood, environmental conservation and cultural preservation are somewhat reconciled. Yet, if ecotourism research continues to remain largely unaware of gender as an analytical category, its contribution to tourism and sustainable development will likely be limited. The adoption of basic, widely employed concepts of gender analysis such as the gender division of labour, access to and control over resources, and levels of empowerment would likely help to rectify this situation. Ultimately, such research may help to develop more gender-aware ecotourism projects, policies and practices, thereby providing increased benefits to both men and women.

References

AITCHISON, C.C. (2005). Feminist and Gender Perspectives in Tourism Studies: The Social-Cultural Nexus of Critical and Cultural Theories. *Tourist Studies* 5(3): 207–224.

AITCHISON, C.C. (Forthcoming). *Gender and Tourism: Social, Cultural and Spatial Perspectives.* London and New York. Routledge.

BELSKY, J. (1999). Misrepresenting Communities: The Politics of Community-based Rural Ecotourism in Gales Point Manatee, Belize. *Rural Sociology* 64: 641–666.

BIANCHI, R.V.(2009). The 'Critical Turn' in Tourism Studies: A Radical Critique. *Tourism Geographies* 11(4): 484–504.

BRANDTH, B. and HAUGEN, M.S. (2007). Gendered Work in Family Farm Tourism. *Journal of Comparative Family Studies* 38(3): 379–393.

BREATHNACH, P., HENRY, M., DREA, S. and O'FLAHTERY, M. (1994). Gender in Irish Tourism Employment. In Kinnaird, V. and Hall, D. (Eds) *Tourism: A Gender Analysis.* New York. John Wiley and Sons: 52–73.

CABEZAS, A.L. (2006). The Eroticization of Labor in Cuba's All-inclusive Resorts: Performing Race, Class and Gender in the New Tourist Economy. *Social Identities* 12(5): 507–521.

CATER, E. (2006). Ecotourism as a Western construct. *Journal of Ecotourism* 5(1/2): 23–39.

CURRIE, D.H. and VERNOOY, R. (2010). Social and Gender Analysis in Development Practice. *Gender, Technology and Development* 14(1): 1–23.

DI CIOMMO, R. C. (2007). Gender, Tourism, and Participatory Appraisals at the Corumbau Marine Extractive Reserve, Brazil. *Human Ecology Review* 14(1): 56–67.

DILLY, B.J. (2003). Gender, Culture, and Ecotourism: Development Policies and Practices in the Guyanese Rain Forest. *Women's Studies Quarterly* 31(3/4): 58–75.

DONOHUE, H. and NEEDHAM, R. (2006). Ecotourism: The Evolving Contemporary Definition. *Journal of Ecotourism* 5(3): 192–210.

DOWLEY, A. (2007). How do Women in Prek Toal Village (Tonle Sap Biosphere Reserve, Cambodia) Participate in and Benefit from Ecotourism? *ISP Collection,* Paper 208. http://digitalcollections.sit.edu/isp_collection/208 – Accessed on 22 May 2011.

EMPOWERING WOMEN OF NEPAL (EWN) (2011). Empowering Women of Nepal.http://www.3sistersadventure.com/EWN/ – Accessed on 22 May 2011.

FAIRBAIRN-DUNLOP, P. (1994). Gender, Culture and Tourism Development in Western Samoa. In Kinnaird, V. and Hall, D. (Eds) *Tourism: A Gender Analysis.* New York. John Wiley and Sons: 121–141.

FENNELL, D. (2002). The Canadian Ecotourist in Costa Rica: Ten Years Down the Road. *Journal of Sustainable Development* 5: 282–295.

FERGUSEN, L. (2010). Analysing the Gender Dimensions of Tourism as a Development Strategy. ICEI Working Paper PP03/09, Universidad Complutense de Madrid. http://www.ucm.es/info/icei/pdf/PP%2003-09.pdf – Accessed on 22 May 2011.

FISH, J., CHICHE, Y., DAY, R., EFA, N., WITT, A., FESSEHAIE, R., JOHNSON, K., GUMISIZIRA, G. and NKANDU, B. (2010). *Mainstreaming Gender into Prevention and Management of Invasive Species.* Washington, DC. Global Invasive Species Programmeme (GISP). http://www.gisp.org/whatsnew/docs/GISP_GenderIASA4.PDF – Accessed on 22 May 2011.

FLACKE-NEUDORFER, C. (2007). Tourism, Gender and Development in the Third World: A Case Study from Northern Laos. *Tourism and Hospitality Planning and Development* 4(2): 135–147.

GENTRY, K. M. (2007). Belizean Women and Tourism Work: Opportunity or Impediment? *Annals of Tourism Research* 34(2): 477–496.

HALL, D., SWAIN, M.B. and KINNAIRD, V. (2003). Tourism and Gender: An Evolving Agenda. *Tourism Recreation Research* 28(2): 7–11.

HE, ZHONGHUA (2003). Forest Management in Mosuo Matrilineal Society, Yunnan, China. In Kelkar, G., Nathan, D. and Walter, P. (Eds) *Patriarchy at Odds: Gender Relations in Forest Societies in Asia.* New Delhi, London and Thousand Oaks, CA. Sage: 147–176.

HIGHAM, J.E.S. (Ed) (2007a). *Critical Issues in Ecotourism: Understanding a Complex Tourism Phenomenon.* Oxford. Butterworth Heinemann.

HIGHAM, J.E.S. (2007b). Ecotourism: Competing and Conflicting Schools of Thought. In Higham, J. (Ed) *Critical Issues in Eco-tourism: Understanding a Complex Tourism Phenomenon.* Oxford. Butterworth Heinemann: 1-19.

HONEY, M. (2008). *Ecotourism and Sustainable Development* (Second Edition). Washington, DC. Island Press.

HONEY, M and STEWART, E. (2002). The Evolution of 'Green' Standards for Tourism. In Honey, M. (Ed) *Ecotourism and Certification.* Washington, DC. Island Press: 33–72.

JACKSON, M. (2010). *Empowering Women of Nepal: An Experience of Empowerment in the Land of the Himalaya.* Unpublished Master's thesis, Prescott College. Prescott, Arizona.

KELKAR, G. (2004). Tourism and Gender Relations in Lijiang, China. In Nathan, D., Kelkar, G. and Walter, P. (Eds) *Globalization and Indigenous Peoples in Asia: Changing the Local-Global Interface.* New Delhi, London and Thousand Oaks, CA. Sage: 225–234.

KELKAR, G. (2009). *Adivasi Women: Engaging with Climate Change.* New Delhi: UNIFEM.http://www.unifem.org.in/economicsecurity.html - Accessed on 22 May 2011.

KELKAR, G. and NATHAN, D (2003). Introduction - Forest Societies in Asia: Gender Relations and Change. In Kelkar, G., Nathan, D. and Walter, P. (Eds) *Patriarchy at Odds: Gender Relations in Forest Societies in Asia.* New Delhi, London and Thousand Oaks, CA. Sage: 13–45.

KELKAR, G., NATHAN, D. and WALTER, P. (Eds) (2003). *Patriarchy at Odds: Gender Relations in Forest Societies in Asia.* New Delhi, London and Thousand Oaks, CA. Sage.

KINNAIRD, V. and HALL, D. (Eds) (1994). *Tourism: A Gender Analysis.* New York. John Wiley and Sons.

KINNAIRD, V. and HALL, D. (1996). Understanding Tourism Processes: A Gender-aware Framework. *Tourism Management* 17(2): 96–102.

KINNAIRD, V. and HALL, D. (2000). Theorizing Gender in Tourism Research. *Tourism Recreation Research* 25(1): 71–84.

KINNAIRD, V., KOTHARI, U. and HALL, D. (1994). Tourism: Gender Perspectives. In Kinnaird, V. and Hall, D. (Eds) *Tourism: A Gender Analysis.* New York. John Wiley and Sons: 1–34.

KUSUJIARTI. S. and TICKAMYER. A. (2000). Gender Division of Labor in Two Javanese Villages. *Gender, Technology and Development* 4(3): 415–439.

LEACH, F. (2003). *Practising Gender Analysis in Education.* Oxford. Oxfam Publishing.

LEÓN, Y.M. (2007). The Impact of Tourism on Rural Livelihoods in the Dominican Republic's Coastal Areas. *Journal of Development Studies* 43(2): 340-359.

LILJA, N. and ASHBY, J.A. (1999). *Types of Gender Analysis in Natural Resource Management and Plant Breeding.* Working Document No. 8. CGIAR Systemwide Programme on Participatory Research and Gender Analysis. http://www.prgaprogramme.org/descargas/impact_assessment / working_documents - Accessed on 22 May 2011.

MARCH, C., SMYTH, I. and MUKHOPADHYAY, M. (1999). *A Guide to Gender-Analysis Frameworks.* London. Oxfam Publishing.

MOLYNEUX, M. (1985). Mobilization without Emancipation? Women's Interests, the State, and Revolution in Nicaragua. *Feminist Studies* 11(2): 227-254.

MOSER, C. (1993). *Gender Planning and Development: Theory, Practice and Training.* New York and London. Routledge.

NAKATA, M. and MOMSEN, J.D. (2010). Gender and Tourism: Gender, Age and Mountain Tourism in Japan. *Malaysian Journal of Society and Space* 6(2): 63-71.

PLENO, M.J.L. (2006). Ecotourism Projects and Women's Empowerment: A Case Study in the Province of Bohol, Philippines. *Forum of International Development Studies* 32: 137-154.

PRITCHARD, A. MORGAN, N., ATELJEVIC, I. and HARRIS, C. (Eds) (2007). *Tourism and Gender: Embodiment, Sensuality and Experience.* Oxford. CABI.

RAZAVI, S. and MILLER, C. (1995). *From WID to GAD: Conceptual Shifts in the Women and Development Discourse (Occasional Paper 1).* Geneva, Switzerland. United Nations Research Institute for Social Development.

SACHS, C.E. (1999). *Gendered Fields: Rural Women, Agriculture and Environment.* Boulder, Colorado. Westview Press.

SCHELLHORN, M. (2010). Development for Whom? Social Justice and the Business of Ecotourism. *Journal of Sustainable Tourism* 18(1): 115-135.

SCHEYVENS, R. (1999). Ecotourism and the Empowerment of Local Communities. *Tourism Management* 20: 245-249.

SCHEYVENS, R. (2000). Promoting Women's Empowerment through Involvement in Ecotourism: Experiences from the Third World. *Journal of Sustainable Tourism* 8(3): 232-249.

SCHEYVENS, R. (2007). Ecotourism and Gender Issues. In Higham, J. (Ed) *Critical Issues in Ecotourism.* Oxford. Butterworth-Heinemann: 185-213.

SEN, A. (1990). Gender and Cooperative Conflicts. In Tinker, I. (Ed) *Persistent Inequalities: Women and World Development.* Oxford. Oxford University Press: 123-149.

STRONZA, A. (2005). Hosts and Hosts: The Anthropology of Community-based Ecotourism in the Peruvian Amazon. *National Association for Practice of Anthropology Bulletin* 23: 170-190.

STRONZA, A. and DURHAM, W.H. (Eds) (2008). *Ecotourism and Conservation in the Americas: Putting Good Intentions to Work.* Wallingford, Oxfordshire. CABI.

SUANSRI, P. (2004). *Community Based Tourism Handbook.* Bangkok, Thailand. REST.

SWAIN, M.B. (1995). Gender in Tourism. *Annals of Tourism Research* 22: 247-266.

SWAIN, B.S. and SWAIN, M.B. (2004). An Ecofeminist Approach to Ecotourism Development. *Tourism Recreation Research* 29(3): 1-6.

TUCKER, H. (2007). Undoing Shame: Tourism and Women's Work in Turkey. *Journal of Tourism and Cultural Change* 5(2): 87–105.

VERNOOY, R. (Ed) (2006). *Social and Gender Analysis in Natural Resource Management: Learning Studies and Lessons from Asia*. Thousand Oaks, California. Sage.

WALTER, P. (2004). Tourism and Forest Management among the Hani in Xishuangbanna, China. In Nathan, D., Kelkar, G. and Walter, P. (Eds) *Globalization and Indigenous Peoples in Asia: Changing the Local-Global Interface.* New Delhi, London and Thousand Oaks, CA. Sage: 207–224.

WEINERT, J.M. (2008). The Construction and Influence of Local Gender Roles on Practice in a Global Industry: Ecotourism in Ecuador. Unpublished Ph.D. Dissertation. Columbus, Ohio. The Ohio State University.

WEAVER, D. (2005). Comprehensive and Minimalist Dimensions of Ecotourism. *Annals of Tourism Research* 32(2): 439–455.

WEAVER, D. and LAWTON, L. (2007). Twenty Years On: The State of Contemporary Ecotourism Research. *Tourism Management* 28: 1168–1179.

ZEPPEL, H. (2006). *Indigenous Ecotourism: Sustainable Development and Management*. Wallingford, UK. CABI.

Part II

Global insights on ecotourism

Experiences and value perceptions of an ecotourism trip – An empirical study of outbound Chinese tourists

Ting (Tina) Li, Fang Liu and Geoffrey N. Soutar

ABSTRACT

Ecotourism is a fast-growing tourism sector, but the value of ecotourism trips from tourists' perspective has seldom been studied, which led to the current study that examined outbound Chinese tourists' value perceptions of ecotourism trips. Data were collected from Chinese tourists on bus tours to a well-known Australian ecotourism site over an one-year period. The study suggested tourists' experiences impacted on perceived ecotour value in a number of ways. In particular, all experience dimensions influenced one or more of the five value dimensions examined, while entertainment, relating, and education experiences influenced each of the five value dimensions. The study offers important implications for both ecotourism researchers and operators attempting to understand and manage the key experiential offerings that could help maximise value to international tourists.

Introduction

Today's consumers desire unique experiences from consumption rather than simply goods and/or services (Pine & Gilmore, 1998). Responding to this trend, academics and practitioners have begun to pay attention to experience; seeing creating consumer experience as an important market strategy (Grewal et al., 2009). Understanding experience is especially important in tourism as 'to consume tourism is to consume experiences' (Sharpley & Stone, 2010, p. 2). Tourism experiences are created from every point of contact at which tourists interact with the resources (e.g. events and activities) destinations offer (Oh et al., 2007). Experience also influences tourist attitude and behaviour (Adhikari & Bhattacharya, 2016). Not surprisingly, therefore, more discussions have been seen in the tourism area about the creation and management of tourist experience (e.g. Aho, 2001; Huang & Hsu, 2010; Kim et al., 2010; Lee et al., 2008; Ryan et al., 2003; Wong & Li, 2015).

Understanding consumers' value perceptions is also vital for any business (Woodruff, 1997), as value is an important indicator of future behaviour (e.g. repurchase and recommendation behaviours) (Parasuraman & Grewal, 2000). Tourists are consumers (Ritchie & Hudson, 2009) and a trip can be regarded as a product. Thus, if tourists perceive value from a trip, they are more likely to revisit a destination and to recommend it to others (Chen & Chen, 2010; Petrick & Backman, 2002). Consequently, there has been increasing interest in examining tourists' value perceptions in order to understand their travel behaviour better (Hallak et al., 2018).

Although both constructs have received increasing attention in recent years, the influence tourism experience has on perceived value is understudied, especially in terms of the differential effects of different types of experience. A good understanding of which tourism experiences influence value perceptions would also improve our understanding of these two important constructs and give insights to tourism operators considering the types of experiences they might help to create.

Such information would be very helpful to ecotourism research and operation, as this sector has a very experiential focus. Ecotourism, which is 'responsible travel to natural areas that conserves the environment, sustains the well-being of local people, and involves interpretation and education' (The International Ecotourism Society, 2019), is one of the fastest growing tourism sectors in the past decade that has a particular focus on environmental conservation and sustainability, perhaps explaining its importance to tourism planners (Kim & Park, 2017). Indeed, while previous ecotourism studies have focussed on its potential benefits to ecology, economy and society (e.g. Clifton & Benson, 2006; Parker & Khare, 2005) and promoting tourists' environmentally responsible behaviour (e.g. Lee & Jan, 2018; Miller et al., 2015; Walker & Moscardo, 2014), few studies have examined the value ecotourism trips provide to tourists and how such the value might be created through tourism experience.

Although most ecotourism studies have suggested that ecotourism contributes to conservation and pro-environmental behaviour, there has been an ongoing debate about the mechanism through which these contributions might be demonstrated (Ballantyne et al., 2011; Buckley, 2009). One of the key measures that could help in this regard is to have a better understanding of the value of ecotourism trips, because value is a primary mechanism to influence behaviour (Parasuraman & Grewal, 2000).

To bridge the gap, the current study examined outbound Chinese tourists, a very important segment for many international tourism markets, including ecotourism. There were over 140 million outbound Chinese tourists in 2017 alone with a total expending of 292 billion US dollars (Statista, 2019), making them the world's largest outbound tourism market and many of them were interested in taking ecotourism trips abroad. More specifically, in 2018, around 23 percent of outbound Chinese tourists reported they had chosen to stay in eco-friendly hotels, and one of their top five preferred activities was taking ecotours (IPSOS, 2018). Not surprisingly, Chinese tourists are a priority target for many ecotourism planners. Although there might be some differences between China and Western countries with regards to some ecotourism perceptions (Buckley et al., 2008; Wen & Ximing, 2008), they agree upon key elements of ecotourism (e.g. nature-based and encouraging pro-environmental behaviour) (Shi et al., 2019; Wang et al., 2012). More recently, there has been a call for research into outbound Chinese tourists, as they might behave differently when travelling overseas (as compared with their domestic trips) (Ma et al., 2018). Consequently, the current study not only provides important insights into the role experience plays in creating value for ecotourism trips, but also improves our understanding of Chinese tourists' ecotourism experience abroad, which can be applied into tourism markets other than ecotourism.

Literature review

Experience

Tourism experience comes from tourists' interactions with a destination's offerings (Walls et al., 2011). Such experiences reflect people's mental states when interacting with these offerings (Otto & Ritchie, 1996). Indeed, 'everything tourists go through at a destination can be experienced, be it behavioural or perceptual, cognitive or emotional, or expressed or implied' (Oh et al., 2007, p. 120).

Pine and Gilmore (1998) suggested a framework with four aspects (entertainment (being entertained by, for example, watching a performance), education (gaining new knowledge), escapism (escaping from daily life and indulging in destination environments), and (a)esthetic (appreciating the beauty of the physical environment)). These so-called 4Es have been used in many tourism studies (e.g. Hosany & Witham, 2010; Loureiro, 2014; Mehmetoglu & Engen, 2011; Oh et al., 2007; Quadri-Felitti & Fiore, 2012; Sidali et al., 2015). Schmitt (1999) suggested an alternative experiential framework with five aspects (sensory (through sight, sound, touch, taste and smell), emotional (inner feelings and emotions such as enjoyment), thinking (cognitive and problem-solving experiences), acting (physical and behavioural experiences such as physical activity) and relating (social experiences with other individuals)) that has also been used in tourism research (e.g. Lee & Chang, 2012; Tsaur et al., 2007), including ecotourism (Wang et al., 2012).

Chan and Baum (2007) suggested ecotourism experiences included: novelty (escape, something new and different), hedonism (excitement and enjoyment) and stimulation (education and challenges) aspects, which are similar to Pine and Gilmore's (1998) aspects. Wang et al. (2012) used Schmitt's (1999) framework and found three ecotourism dimensions (acting, emotion and aesthetics). Some researchers (e.g. Jurowski, 2009) have suggested integrating Pine and Gilmore's and Schmitt's frameworks, which led us to suggest the nine experience dimensions included in this study.

Perceived value

Perceived value is 'a consumer's overall assessment of the utility of a product based on perceptions of what is received and what is given' (Zeithaml, 1988, p. 14). Value is often a multi-dimensional concept (e.g. Liu et al., 2019). The functional aspect of value is seen as the primary driver of buying behaviour (e.g. Chang & Wildt, 1994), but conceptualising perceived value solely on its functional aspect may be inappropriate, as consumers also pay attention to the social and psychological aspects of consumption, especially in services contexts, such as tourism (Holbrook & Hirschman, 1982; Sheth et al., 1991).

Consequently, researchers have developed multidimensional frameworks to explain perceived value (e.g. Babin et al., 1994; Holbrook, 1994; Sheth et al., 1991). Tourism researchers have also addressed the need to examine value in a multi-dimensional way. For example, Sánchez et al. (2006) developed the GLOVAL scale (functional, emotional and social value), while Sweeney and Soutar's (2001) PERVAL framework has been used in an adventure tourism context (functional value, value for money, emotional value, social value, and novelty value) (Williams & Soutar, 2009) and in some recent tourism studies (e.g. Choi et al., 2018).

Most researchers have examined ecotourism trip value from only the value for money perspective (e.g. Castellanos-Verdugo et al., 2016; Chiu et al., 2014; Lee et al., 2013; Oviedo-García et al., 2017; Perera & Vlosky, 2013), although Kim and Park (2017) used the four PERVAL dimensions. Here, Williams and Soutar's (2009) perceived value dimensions were used to measure the value of an ecotourism trip. That is, we included:

- Economic value (or value for money): Ecotourism trips are usually more expensive (e.g. hotels and destinations that minimise carbon emissions charge higher prices) (Hultman et al., 2015). Thus, value for money may be an important consideration.
- Functional value: Although tourists pay premiums for ecotourism trips, they are often not comfortable (Weaver, 2001). For example, ecotourists may use public transport, consume local/recyclable products or dispose of garbage sustainably when on ecotourism trips (Deng & Li, 2015). Some ecotourism-related attributes, such as nature-based sites and interactions with local culture, may help tourists develop functional value perceptions (Chan & Baum, 2007; Jamrozy & Lawonk, 2017).
- Emotional value: Ecotourism trips may arouse positive emotions, such as excitement and relaxation (Kim & Thapa, 2018; Staus & Falk, 2013). Thus, emotional value may be an important dimension of an ecotourism trip.
- Social value: An ecotourism trip is often believed to be a socially responsible product with ecotourists seeing themselves as environmentally and socially responsible (Lee & Jan, 2018). They may perceive social value coming from such a self-identity (Butcher, 2008).
- Novelty value: Ecotourism trips usually deliver novelty value, as such offerings are different from mass tourism (e.g. the environment is usually primitive, activities, such as hiking, are likely to be more challenging and interpretation about the natural environment may satisfy tourists' desire for learning) (Chan & Baum, 2007).

Experience and perceived value

A number of theories suggest consumption experiences influence perceived value. From a means-end perspective, consumption experiences correspond to the attribute level of abstraction (Kerin et al., 1992), while perceived value corresponds to the consequence level of abstraction (Zeithaml, 1988). Thus, experience acts as the 'means' that helps people achieve desired 'ends' (Gutman, 1982). Other theories also help explain the influence experience has on perceived value. The hedonic consumption theory proposed by Hirschman and Holbrook (1982) suggests multi-sensory, emotional, and entertainment aspects influence customers' value evaluations. The relationship between experience and perceived value have been supported by numerous consumer studies (e.g. Babin et al., 1994; Chen & Dubinsky, 2003; Chen & Lin, 2015; Gentile et al., 2007; Kerin et al., 1992).

The impact consumption experiences have on perceived value is also supported by tourism research (e.g. Walls, 2013; Yuan & Wu, 2008), as experiences at a destination seem to be a 'source of value' (Oh et al., 2007, p. 123). A few tourism studies have found different types of experience impact differently on perceived value. For example:

- The economic value (value for money) of a cruise vacation was found to be influenced by emotional experiences (hedonic and novelty) (Duman & Mattila, 2005).
- The economic value of a temple stay was influenced by escapism, education and entertainment experiences, while their emotional value was influenced by entertainment, escape and aesthetic experiences (Song et al., 2015).
- The novelty and emotional value of an adventure trip were influenced by acting experiences (i.e. participation in adventurous activities) (Prebensen & Xie, 2017).

Although a few prior studies have attempted to examine the relationship between ecotourism experience and perceived trip value, their attention was primarily on the value for money, as already mentioned. To develop a more comprehensive understanding of the experience-value relationship in the ecotourism experience context, the current study will investigate this important relationship through a multi-dimension approach, using a group of outbound Chinese tourists who have taken ecotourism trips in Australia.

The present study

The sample

China has become the largest sourcing country for Australia (Tourism Australia, 2018). More than 1.4 million Chinese tourists visiting Australia from July, 2018 to June, 2019 and the number of visitors has increased fourfold since 2009 (ABS, 2019). To capture this important market, the present study adopted a sample of outbound Chinese tourists who had taken one-day package tours from Perth, the capital of Western Australia, to the Pinnacles, one of the most visited WA ecotourism sites. While staying in Perth, Chinese tourists often take different short-trips to well-known sites around Perth, and many of these sites have ecotourism offerings. Most of these tourists were either independent travellers or packaged tourists.

The site

The site selected in the present study is the Pinnacles, which is one of Western Australia's most attracted ecotourism sites (Tourism Western Australia, 2019). Located along the Indian Ocean's Coral Coast in Nambung National Park, the Pinnacles is an ancient desert filled with towering limestone pillars made of seashells that date back millions of years. The tourists in the current study visited the site in small buses. They were provided with guided walks through the desert and rock formations with detailed information provided on the natural environment and its habitats. These tours met the three key requirements of an ecotourism trip (i.e. nature-based, offering environmental education, and ecologically sustainable) (Beaumont, 2001).

Data collection

Data were collected on nineteen different ecotours undertaken by three tour operators between September 2017 and November 2018. The tours were carefully chosen based on similar costs (approximately 150 Australian dollars), length of time (approximately 9 h), route taken (from Perth to the Pinnacles), activities offered (i.e. sightseeing and desert walking) and vehicles used (small tour buses). These tours were particular designed to cater for outbound Chinese tourists (e.g. having Mandarin-speaking tour guides and Mandarin-written information brochures). Several Chinese tourists in these bus tours were Australian citizens or permanent residents, so they were excluded in the data collection. The data were collected almost every month throughout the year to avoid any confounding impact of seasonality.

A self-administered, structured questionnaire was used. One researcher accompanied each of the 19 tour trips in order to ensure the consistency and the quality of the data collection process. While seated on the bus back to Perth (the starting point), each tourist was approached and asked if they would be interested in completing a survey. The researcher was available at the bus to answer any question. A total of 417 questionnaires were returned and, after a preliminary check, 375 useable questionnaires were obtained. Almost every tourist in each tour group participated; suggesting non-response bias was unlikely to be an issue, as more than 85% of those approached provided usable responses.

The measures

The questionnaire asked about respondents' ecotourism experiences, perceived trip value and obtained some background information. The items asking about their entertainment, education, escapism, and aesthetics experiences were taken from Oh et al.'s (2007) experience scale, which was developed from Pine and Gilmore's (1998) framework, while the sensory, emotional, thinking, acting and relating items were adapted from Schmitt's (1999) framework. The perceived trip value items were adapted from Williams and Soutar's (2009) adventure tourism scale, which was based on Sweeney and Soutar's (2001) PERVAL scale. Information was also obtained about respondents' gender, age, education, income and past ecotourism experience. Apart from the background information, all items were measured on a 7-point Likert-type scale that ranged from (1) 'strongly disagree' to (7) 'strongly agree'. The questionnaire was originally designed in English and then translated into Mandarin, using a back-to-back translation technique. A pilot-test in Mandarin among a group of international students who studied at Perth was conducted to ensure the flow and accuracy of the questionnaire.

Analysis and results

The constructs were first assessed for reliability and validity, after which the relationships between ecotourism experiences and perceived trip value were estimated using a PLS-SEM approach. PLS-SEM is a multivariate statistical approach that allows researchers to examine relationships between constructs simultaneously (Hair et al., 2011). PLS-SEM maximises the explained variance of a model's dependent constructs, can analyze reflective and formative indicators and is less impacted by normality and sample size than the covariance SEM approach (Hair et al., 2016; Henseler et al., 2009). PLS-SEM has been used in a number of recent tourism studies (e.g. Agapito et al., 2013; Assaker et al., 2014; Mehmetoglu, 2011; Rasoolimanesh et al., 2016; Song et al., 2011). Here, the WarpPLS version 6.0 programme (Kock, 2017) was used to estimate the models of interest. In this case, the stable3 resampling method with 100 resamples was used to estimate the model, as this approach yields 'compatible and often more precise estimates of the actual standard errors than bootstrapping' (Kock, 2018, p. 8).

Sample profile

The sample had a relatively good gender balance with a slightly larger female representation (55%). In terms of age, around 32% of the respondents were between 18 and 30 years of age, 19% were between 41 and 50, 17% were between 51 and 60, 14% were between 31 and 40, and 12% were 61 or above. The sample had relatively good education backgrounds, with 57% having bachelor's degrees, 21% having master's degrees, and 5% having doctoral degrees. Respondents were also in general from higher income groups (low (13%), low to middle (10%), middle (39%), middle to upper (19%), and high (5%)). Around 35% of the sample reported they had taken ecotourism trips 3–4 times in the past five years, followed by 1–2 times (26%), 5–6 times (13%), 8 times or more (12%), and 6–7 times (5%). The sample appeared to be not only representative of the outbound Chinese tourists to Western Australia, but also of the general profile of outbound Chinese tourists in terms of education, age, income, and gender, according to a recent big-data report based on 130 million outbound Chinese tourists (China Tourism Academy, 2018).

Stage 1- reliability and validity

The constructs were initially examined to ensure they were reliable and had convergent and discriminant validity using the WarpPLS 6.0 programme (Kock, 2017). All of the items related well to their constructs, as loadings were greater than 0.60, which was expected, as all were well-established scales. The scales' means, as shown in Table 1, were all high (ranging from 4.64 (escapism experience) to 6.14 (aesthetic experience)), suggesting respondents generally had very good experiences and saw good value in their tours. However, there was sufficient variation in the scores, as can be seen in the standard deviations, to suggest further analysis would be worthwhile. Not surprisingly, all were negatively skewed, although none were excessively skewed and some were close to being normally distributed (Kock, 2017). The lack of normality in some of the constructs suggested the decision to use a PLS-SEM approach was a sensible one, as such methods are less impacted by non-normality (Hair et al., 2016).

Internal consistency was assessed by computing Composite Reliability (CR) and Cronbach's alpha coefficients. As can also be seen in Table 1, all constructs were reliable, with alpha coefficients ranging upwards from 0.81, while the CR coefficients ranged upward from 0.86. As all were greater than the suggested minimum of 0.70 (Kock, 2017), all were considered reliable. Convergent validity was assessed by computing AVE scores. Fornell and Larcker (1981) argued an AVE score of 0.50 or more implied there was more information than noise, suggesting convergent validity. The AVE scores, which are also shown in Table 1, ranged upward from 0.61, suggesting all of the constructs had convergent validity.

Table 1. Descriptive statistics and measurement properties for the constructs.

Construct	Mean	SD	Cronbach's Alpha	Composite Reliability	AVE Score
Education experience	5.56	0.91	0.87	0.91	0.73
Aesthetic experience	6.14	0.80	0.81	0.91	0.84
Entertainment experience	5.30	1.02	0.91	0.94	0.84
Escapism experience	4.64	1.27	0.88	0.92	0.73
Sensory experience	5.31	0.99	0.86	0.92	0.78
Emotional experience	5.64	0.94	0.88	0.95	0.90
Acting experience	5.80	0.94	0.88	0.93	0.81
Thinking experience	5.07	1.06	0.92	0.94	0.81
Relating experience	5.06	1.09	0.86	0.91	0.78
Functional value	5.73	0.91	0.95	0.95	0.82
Value for money	5.26	1.06	0.94	0.95	0.83
Emotional Value	5.36	0.97	0.92	0.93	0.77
Social value	4.80	1.11	0.94	0.95	0.82
Novelty value	5.28	0.97	0.83	0.86	0.61

Note: All items were measured on a 7-point Likert-type scale that ranged from (1) 'strongly disagree' to (7) 'strongly agree'.

Discriminant validity was initially assessed through Fornell and Larcker's (1981) approach. They suggested that discriminant validity could be assumed if the square root of two constructs AVE scores were greater than their correlation. All of the construct pairs met this requirement, as the highest correlation was 0.70, while the lowest square root of the AVE scores was 0.72. Further, all of the full collinearity VIF scores were less than 3.3, which also suggested discriminant validity (Kock, 2017). Thus, all of the constructs can be used with confidence in the subsequent analysis.

Stage 2- ecotourism experiences and perceived trip value

The relationships between experience and perceived value were examined next. Once again, the WarpPLS programme was used to do this and the results obtained are shown in Table 2, in which the non-significant paths have been excluded to improve readability. As can be seen from the R^2 statistics, tourists' value perceptions are significantly influenced by their experiences, as the R^2 statistics ranged from 0.52 (social value) to 0.70 (emotional value).

Table 2. Paths from the experience dimensions to the perceived value dimensions.

Construct	Functional Value	VFM	Emotional Value	Social Value	Novelty Value
Education experience	0.25 **	0.10 *	0.17 **	0.11 *	0.09 *
Aesthetic experience	ns	0.18 **	0.13 **	0.11 *	0.10 *
Entertainment experience	0.20 **	0.26 **	0.12 *	0.19 **	0.10 *
Escapism experience	ns	ns	ns	ns	0.12 *
Sensory experience	0.11 *	0.18 **	0.20 **	ns	0.14 **
Emotional experience	0.22 **	0.28 **	0.33 **	ns	0.09 *
Acting experience	0.14 **	0.08 *	ns	ns	0.20 **
Thinking experience	0.14 **	ns	0.14 **	0.27 **	0.13 **
Relating experience	0.09 *	0.16 **	0.09 *	0.33 **	0.15 **
R^2 statistic	0.64	0.59	0.70	0.52	0.59

Notes: * Significant at the 5% level; ** Significant at the 1% level.

Further, it is clear from the path coefficients that different experience aspects influence different value aspects differently. For example, respondents' relating experiences influenced their social value perceptions significantly more than they did their functional or emotional value perceptions. Interestingly, respondents' escapism experiences had little impact, influencing only novelty value perceptions and, then, only moderately. Education, entertainment, and relating experiences influenced all value aspects, although somewhat differently. Education experiences influenced functional and emotional value more, while entertainment experiences influenced value for money most and novelty value least. As noted earlier, relating experiences influenced social value most. Acting experiences influenced novelty value most, while, perhaps not surprisingly, emotional experiences influenced emotional value most.

Discussion and conclusions

Experience

The current study examined Chinese tourists' ecotourism experiences using nine dimensions obtained from Pine and Gilmore's (1999) framework and from Schmitt's (1999) framework (i.e. entertainment, education, escapism, aesthetics, sensory, emotional, thinking, acting and relating experiences), as the analyses suggested these nine dimensions had discriminant validity. While aesthetic experiences had the highest mean score (6.14), all of the mean scores were above the midpoint of their scale, suggesting respondents were positive about their trip experiences. The high aesthetic experience score might be explained by the fact that natural scenery is an important part of most ecotourism trips. The good perceptions of emotional experiences indicate that Chinese tourists participated in 'fun' activities, such as hiking, and such activities might develop a positive mood on an ecotourism trip.

Education experiences were rated favourably, perhaps because tourists were provided with environmental education throughout the trip. Entertainment and sensory experiences were not rated highly, suggesting the respondents' ecotourism experiences in this case were less hedonically-focussed, differentiating such experiences from those provided to mass tourists who usually seek pleasure (Urry, 2002). The relating and thinking experiences had lower means, suggesting Chinese tourists were more involved in ecotourism offerings (i.e. scenery, activities, and education) than socialising with other people or thinking by themselves. The escapism experience dimension had the lowest mean score (4.64), suggesting that, although these tourists enjoyed the natural scenery, activities and education, they did not mentally escape their daily life or immerse themselves in a different world (Pine & Gilmore, 1999).

Perceived trip value

The study found Chinese tourists' perceived value of ecotourism trips could be measured as a multidimensional construct with functional, emotional, value for money, novelty and social value aspects, as suggested by Williams and Soutar (2009). Functional value had the highest mean score (5.73), suggesting the tourists interviewed felt they obtained the most utility from the quality and performance of their ecotourism trip. However, all of the value dimensions had scores above the scales' midpoints, suggesting respondents obtained some utility from all aspects. Interestingly, social value had the lowest mean score (4.80), suggesting these tourists obtained less utility from social approval by taking this trip. This is consistent with Lee and Jan's (2018) suggestion that social identity issues may not be as influential as expected in an ecotourism context.

The experience-value relationship

The current study found a strong positive relationship between Chinese tourists' ecotourism experiences and the value they felt they had received from their ecotourism trips. All experience dimensions influenced one or more of the five value dimensions measured and entertainment, relating and education experiences impacted on each of the five value dimensions. Respondents' thinking, emotional, sensory, aesthetic, acting and escapism experiences affected some of the five value dimensions.

Although its mean score was rated relatively low, respondents' entertainment experiences influenced all aspects of trip value significantly. 'Having fun' was valued by these tourists, which was consistent with some prior research (e.g. Packer, 2006), in which 'having fun' was emphasised. The impact of entertainment experience might be explained by the sample. When Chinese tourists travel overseas, they tend to seek hedonic experiences in order to 'liberate the spirit from the mundane [in China]' (Chow & Murphy, 2008; Doctoroff, 2005, p. 57).

Further, although these tourists did not seem to have much time to interact on their ecotourism trip, the relating experience influenced all aspects of trip value. Social interaction is important in Chinese culture (Gabrenya et al., 1996) and is even more important when Chinese people travel to unknown places, as such interactions increase emotional security (Cohen, 1972).

Respondents' education experiences also impacted significantly on all aspects of trip value, especially on its functional value. Interestingly, education is a key aspect of ecotourism (The International Ecotourism Society, 2019). The impact of such experiences suggested these tourists had a good understanding of 'ecotourism' and that they wanted educational experiences when they took ecotourism trips, regarding it as an important indicator of trip quality (i.e. functional value), which differentiates them from mass tourists. This understanding of ecotourism may relate to respondents' educational background and past ecotourism experience (57% had bachelor degrees and 35% had taken ecotourism trips 3 or 4 times in the past five years).

Thinking experiences have been found to impact on perceived social value, which is an interesting finding. This may be because ecotourism trips can generate many self-reflective thoughts (Walter, 2013), which subsequently influence tourists' self-identity evaluation (Taylor et al., 1995). The positive impact of thinking experiences on perceived emotional value can be explained by the widely supported cognition-affect relationship (Smith & Swinyard, 1982). Thinking experiences are also likely to satisfy tourists' curiosity, thus influencing novelty value. However, thinking experiences did not seem to influence the perception of value for money. Past research found that thinking experiences of novice consumers might influence their price evaluations (Monroe, 1973; Sinha & Batra, 1999). The tourists included in the current study are more experienced (i.e. had prior experience of ecotourism trips); therefore, their thinking experiences might not have a significant influence on their value for money perceptions.

Emotional and sensory experiences influenced most of the value dimensions except social value. It is interesting to note that emotional experiences had a significant impact on value for money. This may be because positive emotions generate positive price perceptions, such as price fairness (Campbell, 2007; Ferguson et al., 2011), which is likely to improve the value for money perceptions (Zeithaml, 1988). Although previous studies have mainly identified negative emotions in promoting ecotourism trips (e.g. being sad or angry about environmental problems) (Ballantyne et al., 2011; Staus & Falk, 2013), the current study found it was the positive emotions that contributed most to the perceived value of an ecotourism trip. Further, sensory experiences had a significant impact on emotional value, which is consistent with the findings of previous studies (e.g. Bruner, 1990; Helmefalk & Hultén, 2017; Krishna, 2012; Yoo et al., 1998). However, neither emotional nor sensory experiences influenced perceived social value. As discussed earlier, ecotourism trips can generate self-reflective thoughts and these thoughts might have been largely induced by thinking experiences rather than emotional or sensory experiences.

Acting experiences impacted on three value dimensions, including novelty value, functional value, and value for money. Acting experiences in the current study were mainly reflected by hiking in an ancient desert in Western Australia. Such experiences were new to these outbound Chinese tourists and provided novelty value to them. However, acting experiences did not seem to influence perceived emotional value, which was contradictory to Tsaur et al.'s (2007) study. An explanation might be because acting experiences in the current study (i.e. hiking in a desert) was not as strongly emotion-laden as other activities (i.e. cycling) (Shipway & Stevenson, 2012), and therefore, was unable to provide any significant emotional value. Examining the influence, different types of activities have on perceived emotional value could be an interesting research direction.

Escapism experiences did not influence any value dimensions other than novelty value. Experiences of escaping from daily life and indulging in destination environments, not surprisingly, provide novelty. However, escapism experiences (e.g. feeling like living in a different time or place) involve greater immersion (Pine & Gilmore, 1998) that occurs in an automatic and spontaneous way without any in-depth evaluation (Agarwal & Karahanna, 2000; Csikszentmihalyi, 1990). Immersed people are less discerning and less sensitive to conscious evaluations (Tan et al., 2015).

Practical implications

Tourism operators have recognised the importance of perceived value and see value creation as an important goal. This study shed light into which tourism experiences lead to different types of value in an important market segment (i.e. outbound Chinese tourists) within an ecotourism context. Three tourism experiences (entertainment, relating, and education) were key in creating value, as they influenced each of the five value dimensions and, thus need to be focussed when designing experiential activities and facilities.

Tourism operators need to be aware that entertaining activities affects an ecotour's perceived value most, at least for these Chinese tourists. This suggests tourism operators should offer access to entertaining activities, such as snorkelling, viewing wildlife and hiking, if they wish to deliver trip value to Chinese tourists.

Relating experiences should also be emphasized by tourism operators wishing to increase value perceptions. Travelling with families or friends should be encouraged by giving discounts to group tickets or bookings. Tour guides also play an important role in creating social experiences in ecotourism trips, because their roles as experts of local environment and culture are critical to tourists in remote ecotourism sites (Weaver, 2001). Consequently, guides' professionalism and friendliness are important, suggesting tourism operators should choose guides carefully and need to ensure they are well trained and motivated to provide these aspects well.

Education experiences should be a key, as they improve perceived value, and because it is the essence of ecotourism, whose primary purpose is conservation. Tourism operators should educate tourists not only about the local culture and natural environment which are the knowledge usually provided in ecotourism tours, but should also prompt long-term environmentally responsible behaviour through specific suggestions about reduced consumption on unessential products and reuse of certain materials (i.e. towels rather than tissues) when taking ecotours.

The other six experience dimensions (thinking, emotional, sensory, aesthetic, acting and escapism) should also be promoted, as they influence some value aspects. More specifically, tourism operators should highlight specific experience dimensions if they wish to create specific value dimensions. For example, if generating value for money is a primary concern, operators should pay attention to positive emotional experiences through, for example, providing recreational activities that can help stimulate tourists' pleasure and happy feelings, as emotional experiences contributed the most in improving the value for money perceptions.

Limitations and future research

The current study has some limitations. The single sample (outbound Chinese tourists) and research context (ecotourism trips to the Pinnacles in Australia) may limit the study's generalisability, suggesting future research in other ecotourism destinations and with tourists from other countries is needed. Further, as ecotourism has a particular focus on promoting environmentally responsible behaviours (Litvin, 1996), environmental-related behaviour within or post an ecotourism trip, which might be termed 'environmental value', may be important. As this aspect was not examined here, future studies should address this aspect in ecotourism contexts. In addition, future research should examine some behavioural outcomes of ecotourism trips, such as intention to revisit or recommendation intentions, so as to better understand the perceived value of an ecotourism trip. A couple of recent studies (i.e. Chai-Arayalert, 2020) have discussed the way technology may assist tourists develop positive ecotourism experience. Thus, more attention could be paid to this area. Finally, it is believed that the Covid-19 pandemic may have some long-term impacts on tourism and travel behaviour (Wen et al., 2020) and it would be interesting to understand to what extent tourists' experiences and their value perceptions might have changed during or after the pandemic.

Disclosure statement

No potential conflict of interest was reported by the authors.

Funding

This study was funded by the BHP Billiton-UWA Business School Research Development Award.

ORCID

Fang Liu http://orcid.org/0000-0001-5126-8881
Geoffrey N. Soutar https://orcid.org/0000-0003-1478-788X

References

Adhikari, A., & Bhattacharya, S. (2016). Appraisal of literature on customer experience in tourism sector: Review and framework. *Current Issues in Tourism*, *19*(4), 296–321. https://doi.org/10.1080/13683500.2015.1082538

Agapito, D., Oom do Valle, P., & da Costa Mendes, J. (2013). The cognitive-affective-conative model of destination image: A confirmatory analysis. *Journal of Travel & Tourism Marketing*, *30*(5), 471–481. https://doi.org/10.1080/10548408.2013.803393

Agarwal, R., & Karahanna, E. (2000). Time flies when you're having fun: Cognitive absorption and beliefs about information technology usage. *MIS Quarterly*, *24*(4), 665–694. https://doi.org/10.2307/3250951

Aho, S. K. (2001). Towards a general theory of touristic experiences: Modelling experience process in tourism. *Tourism Review, 56*(3/4), 33–37. https://doi.org/10.1108/eb058368

Assaker, G., Hallak, R., Vinzi, V., & O'Connor, P. (2014). An empirical operationalization of countries' destination competitiveness using partial least squares modeling. *Journal of Travel Research, 53*(1), 26–43. https://doi.org/10.1177/0047287513481275

Australian Bureau of Statistics. (2019). *Australia's Short Term Visitors.* https://www.austrade.gov.au/news/economic-analysis/china-takes-top-spot-in-australias-tourist-rankings.

Babin, B. J., Darden, W. R., & Griffin, M. (1994). Work and/or fun: Measuring hedonic and utilitarian shopping value. *Journal of Consumer Research, 20*(4), 644–656. https://doi.org/10.1086/209376

Ballantyne, R., Packer, J., & Falk, J. (2011). Visitors' learning for environmental sustainability: Testing short-and long-term impacts of wildlife tourism experiences using structural equation modelling. *Tourism Management, 32*(6), 1243–1252. https://doi.org/10.1016/j.tourman.2010.11.003

Beaumont, N. (2001). Ecotourism and the conservation ethic: Recruiting the uninitiated or preaching to the converted? *Journal of Sustainable Tourism, 9*(4), 317–341. https://doi.org/10.1080/09669580108667405

Bruner, G. C. (1990). Music, mood, and marketing. *Journal of Marketing, 54*(4), 94–104. https://doi.org/10.1177/002224299005400408

Buckley, R. (2009). Evaluating the net effects of ecotourism on the environment: A framework, first assessment and future research. *Journal of Sustainable Tourism, 17*(6), 643–672. https://doi.org/10.1080/09669580902999188

Buckley, R., Cater, C., Linsheng, Z., & Chen, T. (2008). Shengtai luyou: Cross-cultural comparison in ecotourism. *Annals of Tourism Research, 35*(4), 945–968. https://doi.org/10.1016/j.annals.2008.07.002

Butcher, J. (2008). Ecotourism as life politics. *Journal of Sustainable Tourism, 16*(3), 315–326. https://doi.org/10.1080/09669580802154116

Campbell, M. C. (2007). "Says who?!" How the source of price information and affect influence perceived price (un)fairness. *Journal of Marketing Research, 44*(2), 261–271. https://doi.org/10.1509/jmkr.44.2.261

Castellanos-Verdugo, M., Vega-Vázquez, M., Oviedo-García, M., & Orgaz-Agüera, F. (2016). The relevance of psychological factors in the ecotourist experience satisfaction through ecotourist site perceived value. *Journal of Cleaner Production, 124*, 226–235. https://doi.org/10.1016/j.jclepro.2016.02.126

Chai-Arayalert, S. (2020). Smart application of learning ecotourism for young eco-tourists. *Cogent Social Sciences, 6*(1), 1–17. https://doi.org/10.1080/23311886.2020.1772558

Chan, J., & Baum, T. (2007). Ecotourists' perception of ecotourism experience in Lower Kinabatangan, Sabah, Malaysia. *Journal of Sustainable Tourism, 15*(5), 574–590. https://doi.org/10.2167/jost679.0

Chang, T., & Wildt, A. (1994). Price, product information, and purchase intention: An empirical study. *Journal of the Academy of Marketing Science, 22*(1), 16–27. https://doi.org/10.1177/0092070394221002

Chen, C., & Chen, F. (2010). Experience quality, perceived value, satisfaction and behavioral intentions for heritage tourists. *Tourism Management, 31*(1), 29–35. https://doi.org/10.1016/j.tourman.2009.02.008

Chen, S., & Lin, C. (2015). The impact of customer experience and perceived value on sustainable social relationship in blogs: An empirical study. *Technological Forecasting & Social Change, 96*, 40–50. https://doi.org/10.1016/j.techfore.2014.11.011

Chen, Z., & Dubinsky, A. (2003). A conceptual model of perceived customer value in e-commerce: A preliminary investigation. *Psychology & Marketing, 20*(4), 323–347. https://doi.org/10.1002/mar.10076

China Tourism Academy. (2018). *Big-data report on outbound Chinese tourists in* 2017. http://www.ctaweb.org/html/2018-2/2018-2-26-11-57-78366.html.

Chiu, Y., Lee, W., & Chen, T. (2014). Environmentally responsible behavior in ecotourism: Antecedents and implications. *Tourism Management, 40*, 321–329. https://doi.org/10.1016/j.tourman.2013.06.013

Choi, M., Law, R., & Heo, C. (2018). An investigation of the perceived value of shopping tourism. *Journal of Travel Research, 57*(7), 962–980. https://doi.org/10.1177/0047287517726170

Chow, I., & Murphy, P. (2008). Travel Activity Preferences of Chinese Outbound Tourists for Overseas Destinations. *Journal of Hospitality & Leisure Marketing, 16*(1-2), 61–80. https://doi.org/10.1080/10507050802096885

Clifton, J., & Benson, A. (2006). Planning for sustainable ecotourism: The case for research ecotourism in developing country destinations. *Journal of Sustainable Tourism, 14*(3), 238–254. https://doi.org/10.1080/09669580608669057

Cohen, E. (1972). Toward a sociology of international tourism. *Social Research*, 39(1), 164–182.

Csikszentmihalyi, M. (1990). *Flow: The psychology of optimal experience*. Harper-Collins.

Deng, J., & Li, J. (2015). Self-identification of ecotourists. *Journal of Sustainable Tourism, 23*(2), 255–279. https://doi.org/10.1080/09669582.2014.934374

Doctoroff, T. (2005). *Billions: Selling to the new Chinese consumer.* Palgrave Macmillan.

Duman, T., & Mattila, A. (2005). The role of affective factors on perceived cruise vacation value. *Tourism Management, 26*(3), 311–323. https://doi.org/10.1016/j.tourman.2003.11.014

Ferguson, J. L., Ellen, P. S., & Piscopo, G. H. (2011). Suspicion and perceptions of price fairness in times of crisis. *Journal of Business Ethics, 98*(2), 331–349. https://doi.org/10.1007/s10551-010-0552-8

Fornell, C., & Larcker, D. F. (1981). Evaluating structural equation models with unobservable variables and measurement error. *Journal of Marketing Research, 18*(1), 39–50. https://doi.org/10.1177/002224378101800104

Gabrenya, W., Jr, K., & Hwang, K. K. (1996). Chinese social interaction: Harmony and hierarchy on the good earth. In M. H. Bond (Ed.), *The handbook of Chinese psychology* (pp. 309–321). Oxford University Press.

Gentile, C., Spiller, N., & Noci, G. (2007). How to sustain the customer experience: An overview of experience components that co-create value with the customer. *European Management Journal, 25*(5), 395–410. https://doi.org/10.1016/j.emj.2007.08.005

Grewal, D., Levy, M., & Kumar, V. (2009). Customer experience management in retailing: An organizing framework. *Journal of Retailing, 85*(1), 1–14. https://doi.org/10.1016/j.jretai.2009.01.001

Gutman, J. (1982). A means-end chain model based on consumer categorization processes. *Journal of Marketing*, *46*(2), 60–72. https://doi.org/10.1177/002224298204600207

Hair, J. F., Hult, G. T. M., Ringle, C. M., & Sarstedt, M. (2016). *A primer on partial least squares structural equation modeling (PLS-SEM)*. Sage.

Hair, J., Ringle, C., & Sarstedt, M. (2011). PLS-SEM: Indeed a silver bullet. *Journal of Marketing Theory and Practice*, *19* (2), 139–152. https://doi.org/10.2753/MTP1069-6679190202

Hallak, R., Assaker, G., & El-Haddad, R. (2018). Re-examining the relationships among perceived quality, value, satisfaction, and destination loyalty: A higher-order structural model. *Journal of Vacation Marketing*, *24*(2), 118–135. https://doi.org/10.1177/1356766717690572

Helmefalk, M., & Hultén, B. (2017). Multi-sensory congruent cues in designing retail store atmosphere: Effects on shoppers' emotions and purchase behavior. *Journal of Retailing and Consumer Services*, *38*, 1–11. https://doi.org/10.1016/j.jretconser.2017.04.007

Henseler, J., Ringle, C. M., & Sinkovics, R. R. (2009). The use of partial least squares path modeling in international marketing. *Advances in International Marketing*, *20*(1), 277–319. https://doi.org/10.1108/S1474-7979(2009)0000020014

Hirschman, E. C., & Holbrook, M. B. (1982). Hedonic consumption: Emerging concepts, methods and propositions. *Journal of Marketing*, *46*(3), 92–101. https://doi.org/10.1177/002224298204600314

Holbrook, M. B. (1994). The nature of customer value: An axiology of services in the consumption experience. *Service Quality: New Directions in Theory and Practice*, *21*(1), 21–71. https://doi.org/10.4135/9781452229102.n2

Holbrook, M. B., & Hirschman, E. C. (1982). The experiential aspects of consumption: Consumer fantasies, feelings, and fun. *Journal of Consumer Research*, *9*(2), 132–140. https://doi.org/10.1086/208906

Hosany, S., & Witham, M. (2010). Dimensions of cruisers' experiences, satisfaction, and intention to recommend. *Journal of Travel Research*, *49*(3), 351–364. https://doi.org/10.1177/0047287509346859

Huang, J., & Hsu, C. H. (2010). The impact of customer-to-customer interaction on cruise experience and vacation satisfaction. *Journal of Travel Research*, *49*(1), 79–92. https://doi.org/10.1177/0047287509336466

Hultman, M., Kazeminia, A., & Ghasemi, V. (2015). Intention to visit and willingness to pay premium for ecotourism: The impact of attitude, materialism, and motivation. *Journal of Business Research*, *68*(9), 1854–1861. https://doi.org/10.1016/j.jbusres.2015.01.013

The International Ecotourism Society. (2019). *What is ecotourism?* http://www.ecotourism.org/what-is-ecotourism.

Ipsos. (2018). *Chinese international travel monitor* 2018. https://www.ipsos.com/en/chinese-international-travel-monitor-2018.

Jamrozy, U., & Lawonk, K. (2017). The multiple dimensions of consumption values in ecotourism. *International Journal of Culture, Tourism and Hospitality Research*, *11*(1), 18–34. https://doi.org/10.1108/IJCTHR-09-2015-0114

Jurowski, C. (2009, July 29–31). An examination of the four realms of tourism experience theory [paper presentation]. *International CHRIE Conference-Refereed Track 23*. University of Massachusetts, Amherst, USA. https://scholarworks.umass.edu/refereed/Sessions/Wednesday/23

Kerin, R., Jain, A., & Howard, D. (1992). Store shopping experience and consumer price-quality-value perceptions. *Journal of Retailing*, *68*(4), 376–397. https://doi.org/10.1177/002224298805200302

Kim, J. H., Ritchie, J. R., & Tung, V. W. S. (2010). The effect of memorable experience on behavioral intentions in tourism: A structural equation modeling approach. *Tourism Analysis*, *15*(6), 637–648. https://doi.org/10.3727/108354210X12904412049776

Kim, K., & Park, D. (2017). Relationships among perceived value, satisfaction, and loyalty: Community-based ecotourism in Korea. *Journal of Travel & Tourism Marketing*, *34*(2), 171–191. https://doi.org/10.1080/10548408.2016.1156609

Kim, M., & Thapa, B. (2018). Perceived value and flow experience: Application in a nature-based tourism context. *Journal of Destination Marketing & Management*, *8*, 373–384. https://doi.org/10.1016/j.jdmm.2017.08.002

Kock, N. (2017). *WarpPLS user manual: Version 6.0*. ScriptWarp Systems.

Kock, N. (2018). Should bootstrapping be used in PLS-SEM? Toward stable *P*-Value calculation methods. *Journal of Applied Structural Equation Modeling*, *2*(1), 1–12.

Krishna, A. (2012). An integrative review of sensory marketing: Engaging the senses to affect perception, judgment and behavior. *Journal of Consumer Psychology*, *22*(3), 332–351. https://doi.org/10.1016/j.jcps.2011.08.003

Lee, S., Chang, S., Hou, J., & Lin, C. (2008). Night market experience and image of temporary residents and foreign visitors. *International Journal of Culture, Tourism and Hospitality Research*, *2*(3), 217–233. https://doi.org/10.1108/17506180810891591

Lee, T., & Chang, Y. (2012). The influence of experiential marketing and activity involvement on the loyalty intentions of wine tourists in Taiwan. *Leisure Studies*, *31*(1), 103–121. https://doi.org/10.1080/02614367.2011.568067

Lee, T., & Jan, F. (2018). Ecotourism behavior of nature-based tourists: An integrative framework. *Journal of Travel Research*, *57*(6), 792–810. https://doi.org/10.1177/0047287517717350

Lee, W., Graefe, A., & Hwang, D. (2013). Willingness to pay for an ecological park experience. *Asia Pacific Journal of Tourism Research*, *18*(3), 288–302. https://doi.org/10.1080/10941665.2011.652970

Litvin, S. (1996). Ecotourism: A study of purchase proclivity. *Journal of Vacation Marketing*, *3*(1), 43–54. https://doi.org/10.1177/135676679600300104

Liu, F., Kanso, A., Zhang, Y., & Olaru, D. (2019). Consumers' evaluations of mobile advertising: A cross-cultural study. *Journal of Promotion Management*, *25*(7), 1028–1058. https://doi.org/10.1080/10496491.2019.1612495

Loureiro, S. (2014). The role of the rural tourism experience economy in place attachment and behavioral intentions. *International Journal of Hospitality Management*, *40*, 1–9. https://doi.org/10.1016/j.ijhm.2014.02.010

Ma, Y., Ooi, C. S., & Hardy, A. (2018). Chinese travelling overseas and their anxieties. In E. C. L. Yang, & C. Khoo-Lattimore (Eds.), *Asian Cultures and contemporary Tourism* (pp. 201–220). Springer.

Mehmetoglu, M. (2011). Examining the relationship between push and pull factors through partial least-squares path modeling. *Advances in Hospitality and Leisure*, *7*, 153–171. https://doi.org/10.1108/S1745-3542(2011)0000007012

Mehmetoglu, M., & Engen, M. (2011). Pine and Gilmore's concept of experience economy and its dimensions: An empirical examination in tourism. *Journal of Quality Assurance in Hospitality and Tourism*, *12*(4), 237–255. https://doi.org/10.1080/1528008X.2011.541847

Miller, D., Merrilees, B., & Coghlan, A. (2015). Sustainable urban tourism: Understanding and developing visitor pro-environmental behaviours. *Journal of Sustainable Tourism*, *23*(1), 26–46. https://doi.org/10.1080/09669582.2014.912219

Monroe, K. B. (1973). Buyers' subjective perceptions of price. *Journal of Marketing Research*, *10*(1), 70–80. https://doi.org/10.1177/002224377301000110

Oh, H., Fiore, A., & Jeoung, M. (2007). Measuring experience economy concepts: Tourism applications. *Journal of Travel Research*, *46*(2), 119–132. https://doi.org/10.1177/0047287507304039

Otto, J., & Ritchie, J. (1996). The service experience in tourism. *Tourism Management*, *17*(3), 165–174. https://doi.org/10.1016/0261-5177(96)00003-9

Oviedo-García, MÁ, Castellanos-Verdugo, M., Vega-Vázquez, M., & Orgaz-Agüera, F. (2017). The Mediating roles of the overall perceived value of the ecotourism site and attitudes towards ecotourism in sustainability through the key relationship ecotourism knowledge-ecotourist satisfaction. *International Journal of Tourism Research*, *19*(2), 203–213. https://doi.org/10.1002/jtr.209

Packer, J. (2006). Learning for fun: The unique contribution of educational leisure experiences. *Curator: The Museum Journal*, *49*(3), 329–344. https://doi.org/10.1111/j.2151-6952.2006.tb00227.x

Parasuraman, A., & Grewal, D. (2000). The impact of technology on the quality-value-loyalty chain: A research agenda. *Journal of the Academy of Marketing Science*, *28*(1), 168–174. https://doi.org/10.1177/0092070300281015

Parker, S., & Khare, A. (2005). Understanding success factors for ensuring sustainability in ecotourism development in Southern Africa. *Journal of Ecotourism*, *4*(1), 32–46. https://doi.org/10.1080/14724040508668436

Perera, P., & Vlosky, R. (2013). How previous visits shape trip quality, perceived value, satisfaction, and future behavioral intentions: The case of forest-based ecotourism in Sri Lanka. *International Journal of Sport Management, Recreation and Tourism*, *11*, 1–24. https://doi.org/10.5199/ijsmart-1791-874X-11a

Petrick, J., & Backman, S. (2002). An examination of the construct of perceived value for the prediction of golf travelers' intentions to revisit. *Journal of Travel Research*, *41*(1), 38–45. https://doi.org/10.1177/004728750204100l005

Pine, B. J., & Gilmore, J. H. (1998). Welcome to the experience economy. *Harvard Business Review*, *76*, 97–105.

Pine, B. J., & Gilmore, J. H. (1999). *The experience economy: Work is theatre & every business a stage*. Harvard Business School Press.

Prebensen, N., & Xie, J. (2017). Efficacy of co-creation and mastering on perceived value and satisfaction in tourists' consumption. *Tourism Management*, *60*, 166–176. https://doi.org/10.1016/j.tourman.2016.12.001

Quadri-Felitti, D. L., & Fiore, A. M. (2012). Experience economy constructs as a framework for understanding wine tourism. *Journal of Vacation Marketing*, *18*(1), 3–15. https://doi.org/10.1177/1356766711432222

Rasoolimanesh, S., Dahalan, N., & Jaafar, M. (2016). Tourists' perceived value and satisfaction in a community-based homestay in the Lenggong Valley world Heritage Site. *Journal of Hospitality and Tourism Management*, *26*, 72–81. https://doi.org/10.1016/j.jhtm.2016.01.005

Ritchie, J., & Hudson, S. (2009). Understanding and meeting the challenges of consumer/tourist experience research. *International Journal of Tourism Research*, *11*(2), 111–126. https://doi.org/10.1002/jtr.721

Ryan, C., Trauer, B., Kave, J., Sharma, A., & Sharma, S. (2003). Backpackers-What is the Peak experience? *Tourism Recreation Research*, *28*(3), 93–98. https://doi.org/10.1080/02508281.2003.11081421

Sánchez, J., Callarisa, L., Rodríguez, R., & Moliner, M. (2006). Perceived value of the purchase of a tourism product. *Tourism Management*, *27*(3), 394–409. https://doi.org/10.1016/j.tourman.2004.11.007

Schmitt, B. (1999). Experiential marketing. *Journal of Marketing Management*, *15*(1-3), 53–67. https://doi.org/10.1362/026725799784870496

Sharpley, R., & Stone, P. R. (eds.). (2010). *Tourist experience: Contemporary perspectives*. Routledge.

Sheth, J., Newman, B., & Gross, B. (1991). Why we buy what we buy: A theory of consumption values. *Journal of Business Research*, *22*(2), 159–170. https://doi.org/10.1016/0148-2963(91)90050-8

Shi, F., Weaver, D., Zhao, Y., Huang, M. F., Tang, C., & Liu, Y. (2019). Toward an ecological civilization: Mass comprehensive ecotourism indications among domestic visitors to a Chinese wetland protected area. *Tourism Management*, *70*, 59–68. https://doi.org/10.1016/j.tourman.2018.07.011

Shipway, R., & Stevenson, N. (2012). Experiencing sport tourism. *Journal of Sport & Tourism*, *17*(2), 81–84. https://doi.org/10.1080/14775085.2012.729897

Sidali, K., Kastenholz, E., & Bianchi, R. (2015). Food tourism, niche markets and products in rural tourism: Combining the intimacy model and the experience economy as a rural development strategy. *Journal of Sustainable Tourism*, *23*(8-9), 1179–1197. https://doi.org/10.1080/09669582.2013.836210

Sinha, I., & Batra, R. (1999). The effect of consumer price consciousness on private label purchase. *International Journal of Research in Marketing*, *16*(3), 237–251. https://doi.org/10.1016/S0167-8116(99)00013-0

Smith, R. E., & Swinyard, W. R. (1982). Information response models: An integrated approach. *Journal of Marketing*, *46*(1), 81–93. https://doi.org/10.1177/002224298204600108

Song, H., Lee, C., Park, J., Hwang, Y., & Reisinger, Y. (2015). The influence of tourist experience on perceived value and satisfaction with temple stays: The experience economy theory. *Journal of Travel & Tourism Marketing*, *32*(4), 401–415. https://doi.org/10.1080/10548408.2014.898606

Song, H., Li, G., van der Veen, R., & Chen, J. (2011). Assessing mainland Chinese tourists' satisfaction with Hong Kong using tourist satisfaction index. *International Journal of Tourism Research*, *13*(1), 82–96. https://doi.org/10.1002/jtr.801

Statista. (2019). *Tourism industry in China-Statistics & Facts*. https://www.statista.com/topics/1210/tourism-industry-in-china/.

Staus, N., & Falk, J. (2013). The role of emotion in ecotourism experiences. In R. Ballantyne, & J. Packer (Eds.), *International Handbook on Ecotourism* (pp. 178–191). Edward Elgar Pub. Ltd.

Sweeney, J. C., & Soutar, G. N. (2001). Consumer perceived value: The development of a multiple item scale. *Journal of Retailing*, *77*(2), 203–220. https://doi.org/10.1016/S0022-4359(01)00041-0

Tan, W. K., Lee, P. W., & Hsu, C. W. (2015). Investigation of temporal dissociation and focused immersion as moderators of satisfaction-continuance intention relationship: Smartphone as an example. *Telematics and Informatics*, *32* (4), 745–754. https://doi.org/10.1016/j.tele.2015.03.007

Taylor, S. E., Neter, E., & Wayment, H. A. (1995). Self-evaluation processes. *Personality and Social Psychology Bulletin*, *21*(12), 1278–1287. https://doi.org/10.1177/01461672952112005

Tourism Australia. (2018). *Traveller snapshot China*. https://www.tourism.australia.com/content/dam/assets/document/1/7/5/c/6/2013126.pdf.

Tourism Western Australia. (2019). *Visitor experience and expectations research (VEER) 2018*-2019. https://www.tourism.wa.gov.au/Publications20Library/Research20and20reports/2019/Ad-hoc/2018-1920VEER%20Report.pdf.

Tsaur, S., Chiu, Y., & Wang, C. (2007). The visitors behavioral consequences of experiential marketing. *Journal of Travel & Tourism Marketing*, *21*(1), 47–64. https://doi.org/10.1300/J073v21n01_04

Urry, J. (2002). *The Tourist Gaze*. Sage.

Walker, K., & Moscardo, G. (2014). Encouraging sustainability beyond the tourist experience: Ecotourism, interpretation and values. *Journal of Sustainable Tourism*, *22*(8), 1175–1196. https://doi.org/10.1080/09669582.2014.918134

Walls, A. (2013). A cross-sectional examination of hotel consumer experience and relative effects on consumer values. *International Journal of Hospitality Management*, *32*, 179–192. https://doi.org/10.1016/j.ijhm.2012.04.009

Walls, A., Okumus, F., Wang, Y., & Kwun, D. (2011). An epistemological view of consumer experiences. *International Journal of Hospitality Management*, *30*(1), 10–21. https://doi.org/10.1016/j.ijhm.2010.03.008

Walter, P. G. (2013). Theorising visitor learning in ecotourism. *Journal of Ecotourism*, *12*(1), 15–32. https://doi.org/10.1080/14724049.2012.742093

Wang, W., Chen, J., Fan, L., & Lu, J. (2012). Tourist experience and Wetland parks: A case of Zhejiang, China. *Annals of Tourism Research*, *39*(4), 1763–1778. https://doi.org/10.1016/j.annals.2012.05.029

Weaver, D. B. (2001). *The encyclopedia of ecotourism*. CABI.

Wen, J., Kozak, M., Yang, S., & Liu, F. (2020). COVID-19: Potential effects on Chinese citizens' lifestyle and travel. *Tourism Review*, ahead-of-print, https://doi.org/10.1108/TR-03-2020-0110

Wen, Y., & Ximing, X. (2008). The differences in ecotourism between China and the West. *Current Issues in Tourism*, *11* (6), 567–586. https://doi.org/10.1080/13683500802475927

Williams, P., & Soutar, G. (2009). Value, satisfaction and behavioral intentions in an adventure tourism context. *Annals of Tourism Research*, *36*(3), 413–438. https://doi.org/10.1016/j.annals.2009.02.002

Wong, I. A., & Li, X. (2015). Destination services and travel experience in the gaming Mecca: The moderating role of gambling as a travel purpose among Chinese tourists. *Journal of Travel & Tourism Marketing*, *32*(1-2), 80–99. https://doi.org/10.1080/10548408.2014.986014

Woodruff, R. B. (1997). Customer value: The next source for competitive advantage. *Journal of the Academy of Marketing Science*, *25*(2), 139–153. https://doi.org/10.1007/BF02894350

Yoo, C., Park, J., & MacInnis, D. J. (1998). Effects of store characteristics and in-store emotional experiences on store attitude. *Journal of Business Research*, *42*(3), 253–263. https://doi.org/10.1016/S0148-2963(97)00122-7

Yuan, Y., & Wu, C. (2008). Relationships among experiential marketing, experiential value, and customer satisfaction. *Journal of Hospitality & Tourism Research*, *32*(3), 387–410. https://doi.org/10.1177/1096348008317392

Zeithaml, V. A. (1988). Consumer perceptions of price, quality, and value: A means-end model and synthesis of evidence. *Journal of Marketing*, *52*(3), 2–22. https://doi.org/10.1177/002224298805200302

Rapid ecotourism development and challenges: The case of ecolodge in Iran

Hamid Reza Irani, Mehdi Golverdi, Raymond Rastegar and Soheil Rezaei

ABSTRACT

In recent years, there has been a significant increase in the number of ecolodges in Iran supported by the government tourism policies. This is also as the country is witnessing a growing interest in popularity of ecotourism trips particularly in northern part of the country. However, such development has also created many issues as the rapid expansion of the ecolodges does not necessarily follow the required guidelines. The purpose of this study is to identify the main problems and challenges of ecolodges in three provinces of Gilan, Golestan and Mazandaran in northern Iran. Qualitative data was collected by interviewing the ecolodge owners in the three provinces. The data analysis led to the formation of 3 main themes and 8 sub-themes. The findings revealed that issues such as the lack of support from the government organizations, weakness in tourism laws and policies, problems related to the geography of ecolodges, lack of knowledge, and also problems related to the local community perceptions are among the problems and challenges ecolodges are facing.

Introduction

Nature-based tourism plays an important role in social and economic development of rural areas particularly in developing countries with relatively undisturbed natural environment. Ecotourism in particular, is a type of nature-based tourism that has become popular in countries with unique natural and cultural attractions. Ecotourism is known as a low impact type of tourism that can conserve the natural environment while improving the livelihoods of local communities (Mgonja et al., 2015). Similarly, de Grosbois and Fennell (2021) referred to ecotourism as an effective tool for protection of local livelihoods, encouraging local empowerment, and a learning opportunity for both local residents and tourists to conserve the natural environment. Many other studies reported that how ecotourism is used as a panacea for development in many areas to address issues such as lack of job opportunities, poverty, and contributing to the local economy (e.g. Fennell & de Grosbois, 2021; Ghaderi et al., 2021; Mafi et al., 2020; Rastegar, 2017). Regardless of how ecotourism is defined, it is usually considered as a sustainable option to save poor rural economies while conserving the natural environment.

However, it should be acknowledged that despite its potential to benefit both community and the environment, ecotourism development still faces many challenges particularly in developing countries. For example, investigating ecotourism and its magnitude in Tanzania, Mgonja reported how challenges such as accessibility problems, inadequate infrastructure, and lack of knowledge for marketing and promotion have negatively affected ecotourism development in the country. A study on governance and local participation in ecotourism in Thailand also highlighted how a lack of local knowledge and participation can challenge the ability of ecotourism to be used as a community-based tool for development (Palmer & Chuamuangphan, 2018). Focusing on the case of an ecolodge on the island Kadavu, Fiji, Mafi et al. (2020), identified ecotourists' satisfaction as an important issue for success of ecotourism. The study demonstrates how ecolodges operators must identify attributes and eco-initiatives to influence ecotourists satisfaction. Similarly, de Grosbois and Fennell (2021) argued how operation of ecolodges plays a crucial factor in determining the success of ecotourism. Analyzing best practices in sixty-five top global ecolodges, the authors discuss how potential challenges are usually overlooked in ecolodge operations. In many areas, limited access to capital required to run an ecotourism business, lack of training or formal education, revenue leakage, lack of government organizations' support and conflict among stakeholders are identified as issues limiting ecotourism success (Mgonja et al., 2015; Rastegar, 2020).

Similar to the global trend, the developing country of Iran has witnessed a significant growth in the number of nature-based travellers particularly in the northern regions with unique natural and cultural attractions (Ghorbanzadeh et al., 2019). Following the increase in the number of tourists coupled with government policy to use tourism as a tool for development in rural areas, there has been a significant growth in the number of ecolodges in the country. Empirical research shows how implementation of conservation programmes in the country has also limited traditional jobs such as agriculture and pushed local people to shift to tourism jobs (Ghaderi et al., 2021). So, the rapid tourism development in many parts of the country with a lack of necessary infrastructure and proper planning has resulted in the creation of many challenges that threaten future growth and sustainability of tourism. For example, in a recent study on Radkan ecolodges in Chenaran city, Iran, Ghasemi and Khoob (2021) identified different strengths and weaknesses for the ecotourism development. They argued that how issues such as the lack of government investment in rural tourism infrastructure and inadequate publicity in relation to ecolodges and tourism attractions threaten the future growth of the ecotourism. In a different study, Rahimi et al. (2019), investigated the challenges ecolodges were facing in the case of Khaf city, Iran. They have identified 40 challenges for ecolodges, out of which 26 factors were identified to significantly affect tourism development. Their findings showed how issues such as the lack of infrastructure, health services, and food quality can negatively affect tourists' perceptions. Hesam (2019) in his research entitled identifying barriers to the establishment and development of ecotourism resorts in rural areas of Gilan province also reported challenges these centres were facing. Interviewing 41 tourism experts, activists, villagers and officials showed that issues such as the land use, marketing, administrative problems, lack of training, facilities and also ineffective management and planning were key obstacles for the establishment and development of ecotourism.

The broader literature shows that various studies have been conducted on ecotourism such as ecotourism and sustainable development (Fennell & de Grosbois, 2021), ecotourism development (Ahmed & Hassan, 2020; Karimi, 2020; Rabiee Mandejin et al., 2021; Salehi et al., 2017), ecolodges and rural development (Anabestani et al., 2018; Ghadiri Masoom et al., 2020; Shojaeifard & Shakoor, 2021), life cycle of ecotourism resorts (Sadin & Pourtaheri, 2020), eco-lodge resort design (Gawad, 2020), ecotourists motivation and satisfaction (Simpson et al., 2020), and the brand of eco-lodge resort (Mic & Eagles, 2019). However, there is a dearth of material that discuss the issues associated with rapid ecotourism development particularly in a developing country context. This issue needs a particular attention as generic western models of nature-based tourism cannot be applied to a local context in a developing country where the issues are embedded within local context and culture (Mkono et al., 2021; Palmer & Chuamuangphan, 2018; Rastegar & Ruhanen, 2021). Recognizing this fact and the need for research, this research note is interested in developing a better understanding of challenges and issues ecolodges are facing while offering useful insights into sustainable ecotourism development in the country.

Study location

Iran has recently witnessed a significant growth in the number of ecolodges. The number of eco-lodge resorts have increased from 400 in 2013 to 2000 in 2019 following the allocation of $ 225 million to tourism development from the National Development Fund (IRNA, 2020). In addition, the government has provided financial assistance from credit institutions and banks to develop ecotourism. For example, in 2019, the Ecotourism Support Fund has allocated 100 billion tomans (equivalent to $ 5 million) for the construction of 1000 ecolodge resorts (Quds online, 2020). Consequently, many investors and the private sector have become more willing to invest in ecotourism development. Such support from the government has been the main reasons for the significant growth in the number of eco-lodges in Iran. The Iranian government has also formulated a law in 2015 entitled, 'The Rules and Regulations of Eco-lodge Concerning the Utilization, Evaluation and Ratings'. This law specifies the steps of setting up eco-lodges and their grading criteria. Accordingly, buildings and facilities, equipment and supplies, guest services, human resources and education, safety and health and accessibility are some of the criteria for establishing and grading ecolodges (MCTH, 2015).

The three provinces of Gilan, Mazandaran, and Golestan in northern Iran are situated between the Caspian Sea and the Alborz-mountains. Such unique environment with dense forests, snow-covered mountains and impressive seashores attracts millions of visitors every year. Based on the report of Statistical Center of Iran (2021), the number of tourists in the spring 2019 for Mazandaran, Gilan and Golestan were 16,161,234, 12,361,109 and 3,642,325 tourists respectively. Mazandaran province with a population of about 3 million people and an area of 23,756 square kilometres is the largest northern province that has 102 ecolodges

(Ramsar Cultural Heritage Office, 2022). Gilan and Golestan provinces each have a population of 2.5 million 2 million respectively and 39 (Gilan Cultural Heritage Office, 2022) and 82 ecolodges (Golestan International Exhibitions Corporation, 2022).

Research methodology

This research employed a purposeful and snowball sampling tools to interview ecolodge owners who had experience in setting up and managing an ecolodge. An interview guide was used to conduct semi-structured interviews harnessing the problems and challenges ecolodge owners were facing. These questions focus on problems in establishing, managing, staffing, marketing, suppling foods, and relationship with other stakeholders. Saturation was reached after conducting 20 interviews in 2020. The sample consisted of 7 female and 13 male participants. On behalf of each ecolodge one person was selected and interviewed. To be representative of the total population, 9 participants from Mazandaran, 4 participants from Gilan and 7 participants from Golestan were selected. All interviews were held in Farsi and then the transcriptions from the interviews were translated to English for data analysis. Then, thematic analysis was employed to analyze the data obtained from the interviews. The data analyses process followed the steps of familiarization, initial coding, sharing identified themes among researchers, modification and labelling the final themes (Braun & Clarke, 2006). In the process, researchers shared their ideas of the initial codes to identify the themes. In the next step, the identified themes and current literature were used to write up the findings and discussion of the study.

Findings and discussion

Data analysis process resulted in identifying three main themes focusing on the key issues and challenges highlighted by the ecolodge owners (Table 1). The findings of this study show how issues such as governance and politics, lack of infrastructure, and sociocultural problems can negatively affect ecotourism development. Investigating such problems is critical particularly as ecotourism has been promoted in the regions as a strategy for sustainable development. The following paragraph shall critically discuss the empirical findings and arguments around each discourse within the space of a research note.

Despite the government efforts to formulate ecotourism development policies in Iran, there are many problems in the implementation phase. The highest level of dissatisfaction of the eco-lodge owners was related to government support and issues such as extremely long and difficult process to receive financial supports. Political and global issues also played important roles which led to a reduction in the number of tourists. The ecolodge owners expressed their concerns by highlighting issues such as economic and political sanctions against Iran, lack of booking system on international reservation websites, instability in political relations, and the COVID-19 pandemic. Other studies also confirmed the existence of similar barriers in the country limiting tourism development (Hesam, 2019) when for example, political instability and lack of government support negatively affected tourism development in the country (Ghasemi & Khoob, 2021).

While the government fiscal policy in recent years have led to an increase in the number of ecolodges, problems related to infrastructure have left business owners facing many challenges affecting the quality of ecotourism services. For example, a significant number of ecolodges in Gilan, Golestan and Mazandaran provinces are located in remote areas with a lacking infrastructure such as access roads, public transportation, water and energy. Such issues can create problems for the ecotourism business particularly in winter and colder months of the year. Similar issues were highlighted in the case of nature-based tourism in Iran when a lack of collaboration between government organizations responsible for tourism development negatively affected tourism development (Rastegar, 2020). Also, the findings of this study revealed that ecolodge owners face financial difficulties in financing the construction of buildings, buying equipment and furniture, paying staff salaries or the cost of energy and water. The findings also revealed that many of the ecolodge owners have not received professional training in areas such as hospitality, advertising, and marketing. Therefore, education and formal training are found to play a key role in operating a successful ecotourism business (Mgonja et al., 2015).

The findings of our study also revealed that a lack of tourists understanding of ecotourism culture, has made their expectations of the facilities and services offered in ecolodge resorts unrealistic. Limited knowledge about local culture can also create cultural conflicts when tourists fail to respect local community and their cultural traditions. On the other hand, the culture and values of tourists can also be different and in some cases conflict with the local values. Ghanbari and Salehi (2014) research confirm this finding and acknowledges that one of the challenges of ecotourism in Iran is a lack of cross-cultural understanding

Table 1. Key themes and findings.

Main themes	Sub-themes	Quotes	Respondents/ experience
Governance and politics	Lack of support from the government agencies	*To get a license, we now have to get more than 25 approvals, all these will take a long time … and you never know what the administrative steps are you need to take.*	P7, Male, more than 6 years
		Advertising by the Cultural Heritage Organization is very limited.	P14, Male, less than 3 years
		Despite all the promises, it is very difficult to get a loan … I have not been successful yet.	P9, Male, less than 3 years
	Political issues	*Our relations with other countries are not very good and it is not good for business.*	P12, Male, more than 6 years
		Due to the economic situation, tourism and travel is becoming something luxury, people do not have much money to spend on tourism.	P20, Male, 6 years
	Weakness in tourism laws and policies	*Administrative processes are constantly changing. What is on the tourism organizations website is very different to what is implemented.*	P18, Female, less than 3 years
		The price of raw materials goes up a lot, but the tariff of the Cultural Heritage Organization has not changed. So, the benefit is very low.	P8, Female, less than 3 years
		Some may get a license without having enough knowledge and skills about ecotourism. Currently, 110 licenses have been issued in Golestan province.	P16, Male, 5 years
Infrastructure problems	Geographical location	*Many tourists bypass our region, we are not on the tourist route.*	P13, Male, 4 years
		Our area is very far from the nearby city and there is no good access road.	P11, Female, less than 3 years
		Our area is all paddy fields. There was no electricity. We paid extra money for electricity transmission line.	P3, Male, less than 3 years
	Financial issues	*I do all the work myself; I do not have enough income to hire an employee … Even financing was very difficult, and we could not complete the work. I sold two of my gold bracelets to set up the ecolodge.*	P4, Female, less than 3 years
		The cost of maintaining a building with traditional architecture is high.	P2, Female, less than 3 years
		Due to the current crises, some tourists pay much less than the tariff.	P19, Male, less than 3 years
		There is no worker here and we have to hire someone from other villages/cities, and they do not usually stay long. Usually our family help.	P1, Male, 5 years
	Lack of knowledge	*I do not know much about advertising … I do not have the skill to use social media platforms.*	P5, Male, less than 3 years
		Finding a traditional architect who knows about these buildings is not very easy.	P15, Male, less than 3 years
		I do not know any foreign language and I cannot communicate well with foreign tourists.	P17, Male, less than 3 years
Sociocultural problems	Tourists' knowledge of ecotourism and eco-lodge	*Some tourists are not aware of ecotourism and Eco lodges culture … some even expect to see hotel facilities.*	P10, Female, 4 years
		Sometimes the guests may dispute with each other or do not follow the rules. On one occasion a guest used a heater in the wooden lodge, and it caught fire … thankfully we quickly managed the situation.	P2, Female, less than 3 years
		Some tourists may dislike the local food and ask for more variety of foods, something they can find in restaurants.	P4, Female, less than 3 years
	Local community acceptance	*Some locals are not happy with the presence of tourists … tourists have different dress code, and it is not culturally accepted by everyone here.*	P6, Female, less than 3 years
		Guests come here to enjoy their time, sometimes make noise late night and neighbours are not happy.	P8, Female, less than 3 years
		Neighbours and the Village Council complained why we do not have enough parking for our guests' cars at the property.	P13, Male, 6 years

between the host and guest. However, this finding contradicts with Kiani et al. (2018) research showing that development of ecotourism fosters a sense of trust and encourages communication between individuals from different cultures.

Conclusion

This research note aimed to investigate the ecotourism challenges in practice with a focus on ecolodges in three popular provinces for tourism in northern Iran. The empirical findings of this work reveal that despite the significant growth and popularity in ecolodges and ecotourism in the regions, the presence of issues such as governance and politics, infrastructure and sociocultural problems can undermine sustainability goals of ecotourism development while threatening its future development. It should be argued that a lack of clear ecotourism guidelines and procedure has created confusion and many issues for the ecotourism practitioners. While government policies encourage ecotourism, the government organizations themselves have been slow in effectively adopting and implementing ecotourism guidelines. Addressing these issues requires a deliberate process to establish an ecotourism committee, consisting of all governmental and non-governmental stakeholders including the practitioners and local communities to encourage a participatory decision making. Such process should provide the opportunity for less powerful stakeholders such as the community and business owners to have a voice by expressing their concerns (Shahi et al., 2022). There is no doubt that approaches such as provision of education and training will increase community participation, reduce conflicts, and promote and protect local culture and natural resources. It is hoped that the findings of this study encourage defining inclusive and well implemented policies to achieve full potential of ecotourism in Iran.

Disclosure statement

No potential conflict of interest was reported by the author(s).

References

Ahmed, M. O., & Hassan, M. S. (2020). Eco lodges promotion in Egypt: Assessing the status Quo. *International Journal of Hospitality & Tourism Systems*, *13*(1), 38–51.

Anabestani, A., Giyahi, H., & Javanshiri, M. (2018). An analysis of the effects of establishing ecotourism residence on development of rural settlements (a case study of Radakan Village, Chenaran County). *Spatial Planning*, *8*(2), 1–24.

Braun, V., & Clarke, V. (2006). Using thematic analysis in psychology. *Qualitative Research in Psychology*, *3*(2), 77–101. https://doi.org/10.1191/1478088706qp063oa

de Grosbois, D., & Fennell, D. A. (2021). Sustainability and ecotourism principles adoption by leading ecolodges: Learning from best practices. *Tourism Recreation Research*. https://doi.org/10.1080/02508281.2021.1875170

Fennell, D. A., & de Grosbois, D. (2021). Communicating sustainability and ecotourism principles by ecolodges: A global analysis. *Tourism Recreation Research*, 1–19. https://doi.org/10.1080/02508281.2021.1920225

Gawad, I. O. (2020). Eco lodge design and architectural education: A new approach for design studios. *International Journal of Engineering Research and Technology*, *13*(11), 3877–3892. https://doi.org/10.37624/IJERT/13.11.2020.3877-3892

Ghaderi, Z., Esfehani, M. H., Fennell, D., & Shahabi, E. (2021). Community participation towards conservation of Touran National Park (TNP): An application of reciprocal altruism theory. *Journal of Ecotourism*, 1–15. https://doi.org/10.1080/14724049.2021.1991934

Ghadiri Masoom, M., Minaei, M., & Darban Astana, A. (2020). The role of ecotourism resorts in the development of rural tourism in rural areas of Khor and Biabank. *Space Economy and Rural Development*, *9*(2), 23–42.

Ghanbari, Y., & Salehi, M. (2014). Ecotourism in Iran and its challenges. *Development Strategy*, *11*(43), 84–102.

Ghasemi, M., & Khoob, S. (2021). Optimal strategies for attracting tourists to rural homesteads using strategic planning tools (case study: Radkan-Chenaran county). *Journal of the Geographical Engineering of Territory*, *5*(9), 15–28.

Ghorbanzadeh, O., Pourmoradian, S., Blaschke, T., & Feizizadeh, B. (2019). Mapping potential nature-based tourism areas by applying GIS-decision making systems in East Azerbaijan Province, Iran. *Journal of Ecotourism*, *18*(3), 261–283. https://doi.org/10.1080/14724049.2019.1597876

Gilan Cultural Heritage Office. (2022). *List of ecotourism resorts in Gilan*. https://www.gilanchto.ir/Site/download?file=1572571410.pdf.

Golestan International Exhibitions Corporation. (2022). *List of ecotourism resorts in Golestan*. http://golestanfair.com/fa/Files/contact/boomgardi.pdf.

Hesam, M. (2019). Identifying obstacles to the establishment and development of ecotourism residence in the rural areas of Guilan Province. *Journal of Rural Research*, *10*(3), 546–559. https://doi.org/10.22059/JRUR.2019.281777.1363

Islamic Republic News Agency. (2020). *2000 units of ecotourism resorts in Iran*. https://www.irna.ir.

Karimi, K. S. (2020). Development of ecotourism, and its role in the development of ecotourism attractions in Aran and Bidgol. *Biannual Journal of Social Studies in Tourism*, *8*(16), 145–168.

Kiani, S., Shaterian, M., & Akbari, S. (2018). Modeling and explaining the consequences of ecotourism tours on improving the indicators of socio-cultural, economic and environmental areas of the village (case study: Damavand County rural areas). *Journal of Socio-Cultural Development Studies*, *6*(3), 81–112.

Mafi, M., Pratt, S., & Trupp, A. (2020). Determining ecotourism satisfaction attributes – A case study of an ecolodge in Fiji. *Journal of Ecotourism*, *19*(4), 304–326. https://doi.org/10.1080/14724049.2019.1698585

Mgonja, J. T., Sirima, A., & Mkumbo, P. J. (2015). A review of ecotourism in Tanzania: Magnitude, challenges, and prospects for sustainability. *Journal of Ecotourism*, *14*(2-3), 264–277. https://doi.org/10.1080/14724049.2015.1114623

Mic, M., & Eagles, P. F. (2019). Cooperative branding for mid-range ecolodges: Costa Rica case study. *Journal of Outdoor Recreation and Tourism*, *25*, 113–121. https://doi.org/10.1016/j.jort.2017.12.001

Ministry of Cultural Heritage, Tourism and Handicrafts (MCTH) Website. (2015). Description of specifications and criteria for construction, operation and grading of eco-lodges in Iran. https://www.mcth.ir.

Mkono, M., Rastegar, R., & Ruhanen, L. (2021). Empowering women to protect wildlife in former hunting tourism zones: A political ecology of Akashinga. *Zimbabwe. Journal of Sustainable Tourism*, 1–21. https://doi.org/10.1080/09669582.2021.1900205

Palmer, N. J., & Chuamuangphan, N. (2018). Governance and local participation in ecotourism: Community-level ecotourism stakeholders in Chiang Rai province, Thailand. *Journal of Ecotourism*, *17*(3), 320–337. https://doi.org/10.1080/14724049.2018.1502248

Quds Online News-Analytical Website. (2020). *Establishment of 2000 ecotourism units in Iran*. https://www.qudsonline.ir/.

Rabiee Mandejin, M., Oroji, H., & Alizadeh, M. (2021). Explaining the development model of ecolodges system in rural areas (case study: Khalkhal county). *Tourism Management Studies*, *16*(53), 359–392. https://doi.org/10.22054/TMS.2021.41163.2119

Rahimi, T., Mansouri Moghadam, S., & Javaherzadeh, F. (2019). Investigating the challenges of tourists in ecotourism resorts (case study: Khaf County). *Annual International Congress of New Findings in Agricultural Sciences and Natural Resources, Environment and Tourism, Tehran*.

Ramsar Cultural Heritage Office. (2022). *List of ecotourism resorts in Mazandaran*. http://mazandchto.ir/service/1365?OrgId=5.

Rastegar, R. (2017). *Environmental protection and local resident attitudes at early stages of tourism development*. The University of Queensland.

Rastegar, R. (2020). Tourism and justice: Rethinking the role of governments. *Annals of Tourism Research*, *85*, 102884. https://doi.org/10.1016/j.annals.2020.102884

Rastegar, R., & Ruhanen, L. (2021). A safe space for local knowledge sharing in sustainable tourism: An organisational justice perspective. *Journal of Sustainable Tourism*, 1–17. https://doi.org/10.1080/09669582.2021.1929261

Sadin, H., & Pourtaheri, M. (2020). Evaluating the life cycle of resorts case study: Turkmen Saharan Golestan Province. *Journal of Spatial Planning*, *24*(2), 201–229.

Salehi, S., Moradi Asl, A., & Abdoli, S. F. (2017). Ecotourism development with an emphasis on local communities' participation (case study: The village of Janat Rudbar, Ramsar County). *Journal of Research and Rural Planning*, *6* (3), 185–201. https://doi.org/10.22067/JRRP.V6I3.62073

Shahi, T., Ziaee, M., Rastegar, R., & Kazemian, G. (2022). *Promoting justice through tourism: The case of tourism development in Qeshm Island*. CAUTHE2022, Brisbane, Australia.

Shojaeifard, A., & Shakoor, A. (2021). Analysis and study of the impact of establishing eco-tourism resorts in Iran on the development of rural areas for sustainable regional development. *Journal of the Geographical Engineering of Territory*, *4* (8), 460–479.

Simpson, G. D., Sumanapala, D. P., Galahitiyawe, N. W., Newsome, D., & Perera, P. (2020). Exploring motivation, satisfaction and revisit intention of ecolodge visitors. *Tourism and Hospitality Management*, *26*(2), 359–379. https://doi.org/10.20867/thm.26.2.5

Statistical Center of Iran. (2021). *Number of tourists in the provinces of Iran*. https://www.amar.org.ir.

Ecotourism in Yancheng Wetland, China

Feifei Xu and Geoffrey Wall

Ecotourism may be a way to protect the resources and promote the local economy in and around wetland areas. Using Yancheng, China, as an example and interviews with key stakeholders and local residents as information sources, it is shown that local residents currently benefit little from tourism and, in consequence, are not strong supporters of the wetland nature reserve. They would welcome more tourism, although they know little about it. Their involvement in tourism is currently limited by their education, skills, lack of capital, cultural traditions, the existing distribution channels and the officials' attitudes toward them. Suggestions are made to increase the benefits accruing to local people in order to enhance their support for the reserve and allow it to be managed in more sustainable ways.

Introduction

It is commonly argued in the western world that local people should be involved in the decisions that affect their lives. Ecotourism is often touted as being a way of protecting resources while promoting local economic development but, to be sustainable, it is important to gain local support for its establishment and operation. Ecotourism is being promoted in many parts of the world where the involvement of local people in decision making over resource allocations is not highly developed and may even be resisted. One such place is China, which declared an 'Ecotourism Year' in 1999, but where the attributes of ecotourism are poorly understood and a top-down decision-making system prevails. It should not be surprising that in such a situation, although there are growing attempts to develop ecotourism, there have been few attempts to examine the views of local people concerning ecotourism and little published literature on the topic emanating from China, although the study by Stone and Wall (2004) of forested areas in Hainan, and the writings of Sofield (http://www.wildlifetourism.org.au/ppts/swtc_sofield_trevor.pdf) and the UNESCO Man and Biosphere programme (Chinese National Committee for MAB 2000) are exceptions. This study addresses attitudes towards and involvement in ecotourism in a large coastal wetland area, a different environmental setting in China from the other studies, and makes recommendations for improvements. The paper considers ecotourism in Yancheng Wetland, Jiangsu Province. Based on interviews conducted in the local community, limitations on the current ability of local residents to benefit from tourism are assessed and the potential for them to gain more benefits is examined in the context of the protection and development opportunities that exist in this huge wetland area. The study is not particularly novel in the questions that are raised, in the research methods that are used, nor even in the findings that are presented. However, it is one of only a limited number of studies that have been published on China and it is concerned with a wetland area that is of considerable international ecological significance, although tourism to the area is currently in an early stage of development.

Before describing the case study, comments will be made on communities and tourism, a brief introduction to the nature of ecotourism will be provided and the characteristics of wetlands will be discussed.

Tourism and Community

According to Joppe (1996), 'Community is self defining in that it is based on a sense of shared purpose and common goals'. Reid (2003) pointed out: 'Community ranges from the smallest geographical unit to large communities of interest, which possess no physical boundaries but rely on common interests among people'. For the purpose of this study, community is defined as a group of people that have common interests and take part in social interaction based on geographical ties. In this study, residents of a village are the focus of attention, although it is also necessary to consider aspects of the county in which the village is located.

Many studies have suggested that community participation should be an important element of tourism development (Murphy 1983; Simmons 1994; Cater and Lowman 1995; Kutay 1992; Britton and Clark 1987; Kangas *et al.* 1995; Oppermann 1993). However, especially in developing countries, the local communities are usually limited in their involvement by their educational levels, skills, and access to capital and even by policies. According to Cevat (2000), there are operational, structural and cultural limits to community participation. Formulating and implementing a participatory tourism development approach usually requires a total change in the socio-political, legal, administrative and economic structures in many developing countries. Consequently, there are usually limitations to the ability of local communities to gain benefits from tourism. This is currently the usual situation in China (Wang and Wall 2005, 2007).

The Nature of Ecotourism

There is now a large literature on ecotourism: a journal (*Journal of Ecotourism*) is devoted to the topic, an encyclopaedia has been published (Weaver 2001) and a variety of books are available on the topic (e.g., Buckley 2004; Fennell 2003). At the same time, there are many definitions of ecotourism and none are universally accepted (Blarney 1995; Buckley 1994; Grams 1995). Although the definition of ecotourism continues to evolve, some features of ecotourism have been widely accepted: (1) tourism activity is based on nature and should be educational, (2) the motivation of tourists is to enjoy the environment, (3) tourism activities should not damage the environment, (4) tourists should respect the culture of local people, and (5) the local community should gain benefits from tourism. Ideally, if these components are in place, then ecotourism should be a sustainable activity. This broad perspective on ecotourism is adopted for this study.

Tourism, including ecotourism, if not well-managed, can damage the environment (Mathieson and Wall 1982; Briassoulis 1992; Liddle 1997). However, ecotourism should provide opportunities to enhance environmental management and reduce damage to the environment (Kangas *et al.* 1995). For example, Warner (1991) indicated that ecotourism protects rainforests in Costa Rica, and Duff (1993) suggested that ecotourism benefits Australian parks. Colvin (1994) described a situation in South Africa in which an ecotourism project was managed by an aboriginal community. They allowed tourists to experience their life and culture but, in order to reduce outside influences, the number and types of tourists were regulated. This project provided income for the local community and also protected the environment. Thus, ecotourism can both harm and be used to protect the environment. The key is appropriate planning and management with local support (Ross and Wall 1999). Nevertheless, it is a challenge to balance the benefits gained from ecotourism and its impacts on environment.

Wetlands

Wetlands are the ecotone between terrestrial and aquatic areas and they are one of the world's most important landscape types (Mistch and Gosselink 1993, Environment Canada 1991). They are an especially important habitat for waterfowl. According to IUCN, wetlands are areas of marsh, fen, peatland or water, whether natural or artificial, permanent or temporary, with water that is static or flowing, fresh, brackish, or salt including areas of marine water, the depth of which at low tide does not exceed 6 metres. China, like many other countries, has adopted this definition and the wetland discussed later in the paper is officially recognized by IUCN following the Ramsar Convention.

Wetlands have traditionally been regarded as mosquito-infested wastelands that possess little or no value unless they are dredged, drained, filled and converted by humans to an urban or agricultural use (Estrin and Swaigen 1993). Today, however, a growing number of people realize that wetlands have physical, biological and socio-economic values. In particular, values related to fish and wildlife, especially waterfowl, have been widely recognized for some time, with awareness of other values growing only in recent years (Mistch and Gosselink 1993).

Ecotourism may be established in and around wetland areas and there is an increasing demand for access to wetlands as recreation and tourism resources. Values and functions of wetlands for recreation and tourism include the following:

(1) Wetlands provide open space for aesthetic enjoyment. Natural wetlands are usually beautiful environments. Williams (1991) pointed out that a current tendency in affluent countries is for population growth to be attracted to amenity coastal environments, including the amenities in wetlands. By 1980, over 118 million people (53 per cent of the US population) lived within 80 km of the coast. In 1940, comparable figures were 60 million (46 per cent of the US population) (Platt 1987).

(2) Wetlands serve as recreation sites for fishing, hunting and observing wildlife and undertaking other recreational activities. As wetlands are the transitional

sites of terrestrial and aquatic systems, they are usually rich in diversity. Every year, more than 60 million people watch migratory birds and 3.2 million people hunt ducks and geese in North America (http://www.ramsar.org/values_recreation_e.htm). Many bird-watching activities are carried out in wetland areas.

(3) Wetlands provide opportunities for education and scientific study. Wetlands offer ideal locations for involving the general public and school children in hands-on learning experiences, in an essentially recreational atmosphere, to raise awareness of environmental issues. A good example is the London Wetland Centre located in the heart of London, UK. It attracts 350,000 visitors and the educational programme inform visitors about the functions and values of wetland ecosystems, biodiversity issues and other environmental matters (http:// www.wwt.org.uk/visit/wetlandcentre).

(4) Benefits from tourism: Many wetlands generate considerable income from tourism and recreational uses to the extent that, in some countries, the revenue is a significant component of the national economy. Patterson (1987) calculated that whereas Ontario's wetlands have an annual worth of around CAD$ 300 million for such products as timber and fish, the much larger sum of CAD$ 800 million is generated as a result of tourism and recreational uses.

As in other environments, wetland recreation may be a factor both in wetland conservation and their destruction (Wall 1998). Williams (1991) showed that wetlands are extraordinarily dynamic ecological systems and, together with other external forces, recreational activities may sometimes deflect or halt primary successional processes. However, recreation use is often one of the strongest economic justifications for wetland conservation. Forces impinging upon wetlands include the following:

(1) Wetland losses through development: Bacon (1987) conducted a case study in the insular Caribbean and pointed out that tourism development impacted severely on wetlands in many islands. Drainage and filling occurred for real estate and resort development and to control biting insects.

(2) Environmental damage: Recreational activities commonly degrade wetland water conditions through drainage modifications and sewage disposal, including water pollution, sedimentation from erosion along paths and trails and, sometimes, new facilities do not harmonize with the environment.

(3) Wildlife losses: Hunting and fishing lead to wildlife losses in many wetlands.

(4) Cultural modifications: It is often the poorest people that depend to the greatest extent on the functions and natural resources of the wetlands, although some argue that traditional cultures have adapted best to the fluctuating ecological cycles. Significant changes or developments upstream affect not only the environmental systems, but also undermine existing resource use systems and the cultures that depend upon them.

Few people usually live in the wetlands. Rather, they live at the edge of the wetlands. There are many instances where wetlands have significant religious, historical, archaeological or other cultural values for local communities, representing a part of both local and national heritage (www.ramsar.org/values_cultural:e.htm). This is a poorly documented function of wetlands.

As wetlands are usually fragile, they may need special protection. Many wetlands are located in areas that are economically backward and the human communities that surround them are often poor and want to get benefits from extracting the wetland resources. On the basis of a survey of local communities in a wetland national park in Greece, Trakolis (2001) suggested that community involvement in decision making is an important condition contributing to wetland protection and the empowerment of local communities is important if wetlands are to be used sustainably (Silvius *et al.* 2000). Thus, ecotourism may be advocated for these areas as it is seen as a way to protect the natural and cultural environments, while promoting the local economy.

The study that follows considers ecotourism in Yancheng wetland in Jiangsu Province, eastern China. Ways to protect the ecosystem of the wetland while enhancing the potential for the local community to benefit from tourism are discussed. However, before the study area is described, the research methods will be presented.

Methodology

Both qualitative and quantitative methods were employed to evaluate the current status of tourism in the wetland, the locals' attitudes towards tourism, factors limiting the benefits gained by the local community and opportunities for local participation in tourism. Following exploratory site visits, fieldwork was undertaken from January to March 2004. Secondary data, including tourism plans, were collected. Primary data included 38 semi-structured interviews conducted with local tourism officials, the managers and staff of the nature reserve, and representatives of stakeholder groups such as the hotel managers and their staff and tour operators. Participants were asked open-ended questions on such topics as the advantages and disadvantages of ecotourism, the potential future development of the area and the benefits currently gained from tourism. Interviews lasted from 20 to 95 minutes. Interviewees were equally distributed by gender and aged between 19 and 48 years. A random household survey was also undertaken in the local community with 200 respondents, representing almost 5 per cent of families. Interviews were conducted face to face with either the male or female head of a family. Both closed and open-ended questions were used and the focus was on the benefits gained from tourism.

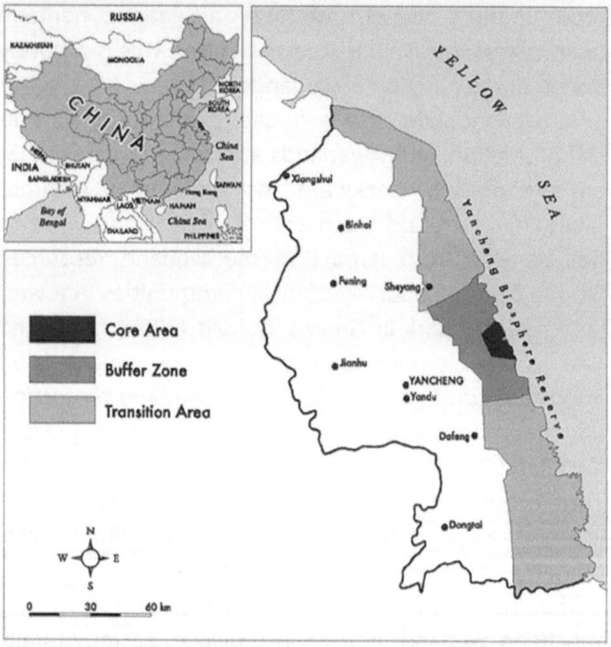

Figure 1. Yancheng, China

Study Area

Yancheng is located in the north of Jiangsu Province in east China (Figure 1). It possesses two national nature reserves: Dafeng, which is famous for Pere David's deer and Yancheng, which is particularly noted for red-crowned cranes. Only the latter is considered here.

The red-crowned crane area is under the jurisdiction of five counties and cities (Dongtai, Dafeng, Sheyang, Binhai and Xiangshui) and has an area of 453,000 hectares (Figure 1). It is the largest coastal wetland reserve in China. It is also rich in species diversity with a recorded total of 450 plants, 379 birds, 45 amphibians, 281 fishes and 47 mammals. Many are on the national list of species requiring protection, e.g., the red-crowned crane, white crane and paddlefish. Yancheng is the important stopover place for migratory birds from Northeast Asia to Australia and it is also an important wintering place for waterfowl. Each spring and summer, approaching 3 million birds fly through this area and every year 200,000 waterfowl winter here (http://sheyang.sina.com.cn/trip/indexl.html). Also, more than 1,000 red-crowned cranes, almost 40 per cent of the world's population, winter in Yancheng. It is the most important winter location for this species. In addition, 29 endangered bird species listed in the IUCN red-data book are found here. Because of its important and rich biodiversity, it became part of the World Biosphere Protection Network in 1992 and in 1999 it became a member of the 'Northeast Asia - Australia Migration Bird Protection Network'.

Yancheng Wetland is on the coast. The Yangtze and Yellow rivers bring massive amounts of sand and sediments to this area so that the wetland extends seawards by approximately 3,333 ha annually. Each year the newly- developed mud flat is separated artificially from the sea by a dyke. At the same time, human uses eat away at the landward side of the wetland. The wetland is in the process of conversion from a saltwater to a freshwater system. The vegetation is mostly composed of reeds and aquatic products are gathered from the mud flats.

In 2000, there were 4,869 families in the county and 85 per cent of workers were employed as farmers, fishers and herdsmen (Sheyang Statistical Bureau 2000). The nearest village to the nature reserve is called 'Xin Yanggang'. It is approximately 4 km from the nature reserve. The population is about 8,000 people and 6,000 depend on farming. It used to have the status of a town but, in 2000, it was incorporated into Huang Shagang town, which is 22 km away, and the local government was also moved to Huang Shagang. Transference of the administration to Huang Shagang caused the village community to become concerned about the local government and its future development.

As tourism is at an early stage, there are very few tourism activities in the area. They are essentially restricted to bird watching and sightseeing. The annual income of the nature reserve is 400,000 yuan (US$ 50,000 approx.), which mostly comes from tickets sales. The present ticket price is 20 yuan (US$ 2.50 approx.). Approximately 30,000 tourists visit the area annually but few facilities and activities are available. The existing hotel is contracted to immigrants from Huang Shagang. It offers 70 beds and is virtually sufficient to meet peak demands for most tourists do not stay overnight.

There are some problems caused by human activities, such as the gathering of reeds for fuel, draining of coastal areas for agricultural and aquaculture, and catching of fish, shrimp, crabs and clams. Furthermore, although ecologically significant, the Yancheng wetland lacks scenic diversity. The red-crowned cranes are concentrated in the core area of this very large wetland so it is not easy for tourists to see them. Nevertheless, Yancheng could have strong tourism potential if developed and marketed appropriately.

Findings

Characteristics of Respondents

Surveys were completed with 200 randomly selected households. Respondents ranged in age from 18 to 78, with 127 male (almost 63.5%) and 73 female participants. The education levels of respondents were generally low: 20 per cent had only studied at primary school, 25 per cent had graduated from junior middle school, 45 per cent had graduated from senior middle school, and only 10 per cent had graduated from a college or university.

For the 38 stakeholder interviews, interviewees were equally distributed by gender and aged between 19 and 48 years. Their education levels were higher than those in the household survey: 8 per cent had only studied at primary school, 18 per cent had completed junior middle school, 53 per cent had passed senior middle school, and 21 per cent had graduated from college or university.

Attitudes Towards the Reserve

Of the 200 respondents to the household survey who were asked whether they knew why the nature reserve was set up, 20 per cent did not know. However, many recognized that it was established to protect the environment and, especially, the red-crowned crane: 20 per cent indicated that it was set up to protect the environment, 20 per cent to protect wildlife, 10 per cent to protect the red-crowned crane as well as other wild animals, and 5 per cent to protect the red- crowned crane specifically. Thus, there was widespread awareness that the reserve was put in place to protect the red-crowned crane and its habitat. Most (89 per cent) of the respondents believe that it is effective in protecting the birds, as illegal hunting has been reduced. The officials think that it has a very good effect on protecting the birds and other endangered species and in promoting local people's awareness of the need for environmental protection. But there is still some illegal hunting, usually by local people.

Before the reserve was set up, almost all (95 per cent) local people depended on fishing in the ocean and on the mudflats. When the reserve was set up, they had to become farmers. However, the 362 ha of cultivable land is insufficient to support the community and young people have migrated to cities such as Yancheng, Shanghai and Nanjing to find jobs; almost 400 people are labourers who work outside of Xin Yanggang.

When the nature reserve was established, no compensation was made to the local people. Almost all (96 per cent) of respondents think that they should get compensation from the government. Also, wildlife from the nature reserve sometimes eats crops and local respondents believe that they should be compensated for the damage. The managers (all but one) of the nature reserve did not think that it was necessary to compensate local people for loss of use of resources, as the land is owned by the government and the use right of this land belongs to the Forestry Centre, which operates as a collective. The villagers disagree. They believe that although the land is occupied by the Forest Centre, they should be allowed to do some fishing, to catch some birds, and also to farm land as it is a collective. But when the nature reserve was set up, they were not allowed to do such things. Thus, there are considerable differences in attitudes between officials and local people.

While the staff of the nature reserve thinks that the relationship with the local community is good, the local people do not like the reserve and they distrust the government: 70 per cent of local respondents think that the current management is not good and that the future is uncertain with such management. The salary of reserve staff is much higher than most people in the local community. The average income of a local family is about 10,000 yuan (US$ 1,250 approx.) per year, or almost 300 yuan (US$ 37.50) per person per month, while the salary of an ordinary staff member of the nature reserve is between 1,000 (US$ 125 approx.) and 1,500 yuan (US$ 187.50 approx.) per month. There is a huge difference in income between local people and reserve staff and, in consequence, local people admire those who are engaged in the tourism industry.

The importance of red-crowned cranes is widely acknowledged. Although poor, local people indicated that they are willing to pay to protect the red-crowned crane: 15 per cent would be willing to pay more than 100 yuan (US$ 12.50), 10% would pay 80-100 yuan(US$10-12.5), 20 per cent would pay 30-80 yuan (US$ 3.75 - 10.00), 35 per cent would pay less than 20 yuan (US$ 2.50) and only 20 per cent would pay nothing. However, many emphasized that if the money were to go to the nature reserve, then they would pay nothing. However, if they were sure that the money would really go to protecting the red-crowned crane and other wildlife, then they would pay. This further shows that the nature reserve does not have a good relationship with the local people. Furthermore, only if the locals can benefit from the nature reserve, are they likely to really protect it; otherwise it is only a passive acceptance, not an active protection.

Tourism and the Community

There was little awareness of or familiarity with ecotourism on the part of household survey respondents. When asked about ecotourism, 50 per cent indicated that they had never heard of it and 15 per cent had heard about it but did not know what it is. Only 35 per cent could say something about ecotourism but their knowledge was partial at best: 71.4 per cent of these respondents (N= 70) thought that ecotourism protected nature, 14.2 per cent thought it meant no pollution and 14.4 per cent thought it was to love living creatures. Local people did not realize that they themselves are also part of the tourism resource and they did not expect to benefit from tourism as a right. In contrast, all of the officials could comment on ecotourism. Almost two- thirds (62 per cent) could address environmental aspects of ecotourism.

When asked about the impacts of tourism, almost all (95 per cent) local people did not expect tourism to have any negative social impacts. Only a few (5 per cent) said that there might be negative impacts to the local community, but half of these could not specify what these impacts might be. Most of the remainder mentioned the possibility of reduced access to natural resources. Individually, all respondents regarded tourism positively, thinking that tourism would bring economic benefits and would enhance their knowledge. They did not expect any adverse impacts to occur to themselves. As for the representatives of the stakeholder groups, while some of them (37 per cent) know that there could be negative impacts of tourism to their community, these are minimized for the sake of their business interests.

The above results indicate that local people have only a vague impression of tourism and what it might bring to them. They also show that the villagers lack information and education on ecotourism. The education levels of respondents were generally low: 20 per cent had only studied at primary school, 25 per cent had graduated from junior middle school, 45 per cent had graduated from senior middle school, and only 10 per cent had graduated from university or college. Even if tourism is developed, they do not have the skills to engage in tourism-related activities and this will limit the benefits that they can derive from tourism. However, 38 per cent of officials felt that the community should have no role in tourism development. Their responses have some similarities with the early academic definitions of ecotourism which focused primarily on environment, cultural and local development factors being added later. Thus, the local residents will require assistance if they are to participate in tourism, and the officials also require education, albeit of a different kind!

With respect to community benefits from tourism, the most widely recognized current benefit according to the survey respondents is improved infrastructure: 74 per cent mentioned that the road had been widened and 21 per cent mentioned improved electricity supply. Only 5% were unable to mention a current benefit. Local benefits are small at present and the reasons for this, according to local respondents, are lack of policy (33 per cent), lack of talented people (26 per cent), lack of money (17 per cent), poor road conditions (11 per cent), lack of good management (8 per cent) and cultural limitations (5 per cent). The existing distribution system limits community benefits. As the nature reserve is managed by the Jiangsu Provincial Environmental Protection Bureau, the financial benefits accrue to the Bureau and not to Sheyang local government, nor to Yancheng municipal government. The local government and Xin Yanggang have no access to these benefits. As a consequence, the government of Sheyang County does not like to promote tourism to the nature reserve because they think that they gain nothing.

According to many researchers (Ross and Wall 1999; Rothman 1978; Perdue 1990; Wearing and Neil 2000), one of the most important local benefits of ecotourism is that it can provide employment opportunities. In this nature reserve, only 20 per cent of the employees are local people from Xin Yanggang and Huang Shagang town. Others come from other parts of Yancheng, or even from elsewhere. The local government is now thinking of trying to attract foreign investors. Depending upon how this is done, it could stimulate local employment or result in substantial leakages. Local people would welcome industrial development because they believe that it could provide them with more opportunities than tourism. In fact, when the nature reserve was established, factories in Xin Yanggang were moved to other areas, followed by the relocation of the administrative centre, leaving the village to become more and more stagnant.

Future Investment and Involvement

One of the problems for management in the reserve is lack of funding - national status does not mean national funding. The survey reported above shows that some funding might be available from one-time local donations. Most (83 per cent) respondents would like to participate in future investments in tourism activities, if policy and personal circumstances allow. Only 13 per cent would choose to farm and 4 per cent did not know. Among those who would like to participate in tourism, 55 per cent would like to run a restaurant, 50 per cent wanted to run a souvenir shop, 45 per cent were interested in providing recreational programmes such as fishing, boating and parking, 35 per cent would like to run a small shop to sell soft drinks, 30 per cent would like to run a cottage or a small inn and 20 per cent would like to have shares in the nature reserve (respondents could select more than one choice). Almost half (45 per cent) would like initiatives to be rim by individuals, 30 per cent by joint ventures with the government and 25 per cent would welcome foreign investment. These data show that even though local people do not know much about tourism, they have an interest in investment and participation but the availability of the resources for them to do so is another matter. As mentioned above, many local people distrust the government but most are unable to bear the risk of investment. However, government representatives did not believe that they could attract foreign investment because of the less developed economy, location and poor transportation links.

The local stakeholders are eager to promote tourism activities. However, an existing store which sells souvenirs and other gifts to tourists is owned by the nature reserve management office and the staff work there on a fixed salary. Two nearby museums also sell gifts but are operated by contractors from outside of Yancheng.

Conclusions and Recommendations

Based on the information presented above, some conclusions can be made:

(1) As tourism is at an early stage, social and economic impacts on the local community are currently limited.

(2) Local people gain few benefits from tourism because their participation is limited by their educational levels, lack of entrepreneurial skills, traditions, officials' attitudes towards local involvement and external domination of the existing distribution channels.

(3) As a result, the locals do not like the nature reserve. It has displaced them from their traditional way of living. However, most would welcome tourism and they express willingness to become involved in many ways.

(4) The reserve lacks money for development. There are few tourism facilities and little interpretation.

Because there are few facilities and the birds are difficult to see, tourists are often disappointed with the experiences available at the nature reserve. At the same time, local people are disappointed with the present development status and the prospects for future development. The following suggestions have been made to the nature reserve management and the local city, Yancheng to alleviate the problems:

(1) Build an interpretation centre on one of the dykes. The life of the red-crowned crane, its habitat and ecological significance, and its role in Chinese culture should be interpreted. Visitors should be introduced to different wetland types, such as salt, brackish and freshwater marsh. Some trees should be planted near the interpretation centre to provide shade, to diversify the landscape and to attract birds to locations visible to visitors. Care should be taken not to introduce alien species.

(2) Limited recreational and tourism activities should be carried out. Hides and boardwalks could be built to facilitate wildlife viewing. Hiking and horseback riding could be encouraged in the forest areas to diversify the product and extend length of stay. Boat trips into the wetland could be arranged for bird watchers and photographers.

(3) Local people should be allowed to do some fishing and breed aquatic species in the buffer areas (it is a large reserve that can be zoned for different uses).

(4) Education programmes should be provided for local people concerning the nature of ecotourism, its potential benefits, and the value of protecting the environment. Training opportunities should be provided for locals in such activities as guiding, photography, running small tourism businesses and craft-making.

(5) The reserve could be developed as a reception centre, resulting in jobs for local people. Only the interpretation centre should be built in the reserve and even that may be better placed outside the reserve. Other facilities such as restaurants, tea houses and cabins should be located in the reserve and built by local investors. However, they should be planned by a local committee in order to avoid haphazard construction.

(5) Small loans should be made available to locals to facilitate investment in tourism and international funding should be sought for protection of the wetland.

(6) The distribution system should be reformed to ensure that the nature reserve and the local community share in the benefits.

(7) A plan for the community should be initiated. Although few local people have much formal education or know much about tourism, they know the area and they understand their own needs.

Although it may not surprise most readers, it may surprise local officials that community members have knowledge and insights to share when allowed to do so, and even skills (such as in boating) that could be used to enhance the ecotourism product. Education and training of both residents and local officials are required and both the reserve itself and local residents require greater access to capital if they are to be able to establish competitive ecotourism products.

It should not be assumed that ecotourism will automatically benefit local communities and generate money to protect the environment. Plans and management strategies are needed for both the reserve and the local community. Even then, it is unlikely that all local people will benefit directly and immediately. Once the local community has gained some benefits from tourism, they will be more likely to support and protect the resources on which it depends.

More generally, through use of a mixed-method research approach, data were gathered and have been reported on the views of local residents concerning ecotourism in a context in which their opinions are seldom heard. Although some findings are specific to Yancheng and to wetlands, the general situation that is described is not unusual and can be found in the majority of nature reserves in China and perhaps many other developing counties.

References

Bacon, P. R. (1987). Use of Wetlands for Tourism in the Insular Caribbean. *Annals of Tourism Research* 14(1): 104-117.

Blamey, R. K. (1995). *The Nature of Ecotourism.* Occasional paper No 21. Canberra. Bureau of Tourism Research.

Briassoulis, H. (1992). Environmental Impacts of Tourism: A Framework for Analysis and Evaluation. In Briassoulis, H. and Straaten, V. D. (eds) *Tourism and the Environment: Regional, Economic and Policy Issues.* Dordrecht. Kluwer Academic Publishers: 79-83.

Britton, S. and Clark, W. C. (eds) (1987). *Ambiguous Alternative: Tourism in Small Developing Countries.* Suva, Fiji. University of the South Pacific.

Buckley, R. C. (1994). A Framework for Ecotourism. *Annals of Tourism Research* 21(3): 661-669.

Buckley, R. C. (2004). *Environmental Impacts of Ecotourism.* Oxon. CAB International.

Cater, E. and Lowman, G. (eds) (1995). *Ecotourism: A Sustainable Option?* Chichester. Wiley and Sons.

Cevat, T. (2000). Limits to Community Participation in the Tourism Development Process in Developing Countries. *Tourism Management* 21(6): 613-633.

Chinese National Committee for Mab (2000). *Study on Sustainable Management Policy for China's Nature Reserves.* Beijing. Chinese National Committee for MAB.

Colvin, J. G. (1994). Ecotourism: A Sustainable Alternative. *NACLE Report on the Americas,* 28, 9

Duff, L. (1993). Ecotourism in National Parks: Impacts and Benefits. *National Parks Journal* 37: 18-20.

Environment Canada (1991). *The State of Canada's Environment.* Ottawa. Environment Canada.

Estrin, D. and Swaigen, J. (1993). *Environment on Trial: A Guide to Ontario Environmental Law and Policy* (Third edition). Toronto. Canadian Institute for Environmental Law and Policy and Emond Montgomery Publications.

Fennell, D. (2003). *Ecotourism: An Introduction.* New York. Routledge.

Joppe, M. (1996). Sustainable Community Tourism Development Revisited. *Tourism Management* 17(7): 475-479.

Kangas, P., Shave, M. and Shave, P. (1995). Economics of an Ecotourism Operation in Belize. *Environmental Management* 19: 669-673.

Kutay, K. (1992). Ecotourism Marketing: Capturing the Demand for Special Interest Nature and Culture Tourism to Support Conservation and Sustainable Development. Paper presented to the third inter-American congress on tourism. Cancun, Mexico.

Liddle, M. (1997). *Recreation Ecology.* London. Chapman and Hall.

Mathieson, A. and Wall, G. (1982). *Tourism: Economic, Physical and Social Impacts.* Harlow. Longman.

Mitsch, W. J. and Gcosselink, J.G. (2000). *Wetlands* (Third edition). NewYork. John Wiley and Sons.

Murphy, P. E. (1983). Community Attitudes to Tourism: A Comparative Analysis. *Tourism Management* 3(2): 189-195.

Oppermann, M. (1993). Tourism Space in Developing Countries. *Annals of Tourism Research* 20(3): 535-556.

Orams, M. B. (1995). Towards a More Sustainable Form of Tourism. *Tourism Management* 16(1): 3-8.

Patterson, N. (1987). Toward a Wetland Protection Strategy for Ontario. *National Wetlands Newsletter* 9: 16-17.

Perdue, R., Long, P. and Allen, L. (1990). Resident Support for Tourism Development. *Annals of Tourism Research* 17(4): 586-599.

Platt, R. H. (1987). Overview of Developed Coastal Barriers. In Platt, R. H., Pelczarski, S. G. and Burbank, K. R. (eds) *Cities on the Beach: Management Issues of Developed Coastal Barriers.* Department of Geography Research Paper No. 224. Chicago. University of Chicago.

Reid, D. G. (2003). *Tourism, Globalization and Development: Responsible Tourism Planning.* London. Pluto Press.

Ross, S. and WALL, G. (1999). Ecotourism: Towards Congruence between Theory and Practice. *Tourism Management* 20(1): 123-132. Rothman, R. (1978). Residents and Transients: Community Reaction to Seasonal Visitors. *Journal of Travel Research* 16: 8-13.

Sheyang Statistical Bureau (2000). *Annual Accounts for Sheyang 1999.* Sheyang. Sheyang Publication House of Statistics

Silvius, M., Oneka, M. and Verhagen, A. (2000). Wetlands: Lifeline for People at the Edge. *Physics and Chemistry of the Earth, Part B: Hydrology, Oceans and Atmosphere* 25(7-8): 645-652.

Simmons, D. G. (1994). Community Participation in Tourism Planning. *Tourism Management* 15(2): 98-108.

Smith, W. J. and Gosselin K, J. G. (2000). *Wetlands* (Third edition). NewYork. John Wiley and Sons Inc.

Stone, M. and Wall, G. (2004). Ecotourism and Community Development: Case Studies from Hainan, China. *Environmental Management* 33(1): 2-24.

Trakolis, D. (2001). Local People's Perceptions of Planning and Management Issues in Prespes Lakes National Park, Greece. *Journal of Environmental Management* 61: 227-241.

Wall, G. (1998). Implication of Global Climate Change for Tourism and Recreation in Wetland Areas. *Climate Change* 40(2): 371-389.

Wang, Y. and Wall, G. (2005). Resorts and Residents: Stress and Conservatism in a Displaced Community. *Tourism Analysis* 10(1): 37-53.

Wang, Y. and Wall, G. (2007). Administrative Arrangements and Displacement Compensation in Top-down Tourism Planning: A Case from Hainan Province, China. *Tourism Management* 28(1): 70-82.

Warner, E. (1991). Ecotourism: New Hope for Rainforests? *American Forests* March/April: 37-44

Wearing, S. and Neil, J. (2000). *Ecotourism: Impacts, Potentials and Possibilities.* Oxford. Reed Education and Professional Publishing.

Weaver, D. (2001). *Encyclopedia of Ecotourism.* Oxford. CAB International.

Williams, M. (1991). *Wetlands: A Threatened Landscape.* Oxford. Basil Blackwell.

Websites Visited

http://www.ramsar.org/values_ recreation_e.htm - Accessed on 7 August 2004.

http: //www.ramsar.org/values_cultural:e.htm - Accessed on 7 August 2004. http://www.wwt.org.uk/visit/ wetlandcentre - Accessed on 5 June 2004.

http://www.ramsar.org/values_ recreation_e.htm - Accessed on 7 August 2004. http://www.wildlifetourism.org.au/ppts/ swtc_sofield_trevor.pdf - Accessed on 22 July 2005. http:// sheyang.sina.com.cn/trip/indexl.html - Accessed on 15 January 2007.

Ecotourism in Gangotri region of the Garhwal Himalayas

Gitanjali Chaturvedi

The Gangotri region in the Garhwal Himalayas attracts a large number of visitors each year: pilgrims, mountaineers, trekkers and sightseers. In addition, many people come to the region in search of jobs as guides, porters, restaurant workers and so on. The region is a unique mix of diverse geographical, demographic and social characteristics, heightened by an intensified sense of religiosity. The area has served as a traditional trade and pilgrim route. With time, development and modernisation, the number of people visiting the area has multiplied. In the light of such heightened activity in a region that is primarily considered sacred, tourism serves as the chief livelihood of the people. Given the diversity of tourists, conflicts are bound to arise between various activities, among the different categories of visitors. In addition, these activities also have a lasting impact on the environment. The paper analyses the conflicts that emerge due to the activity: between tradition and modernity, culture and environment and finally, between environment and employment. The paper examines the impacts of tourism on culture and tradition and the transformation within the same. An analysis on the changing nature of Ashrams as being hermitages for the recluse to providing shelter to pilgrims and operating as micro-economic units is also undertaken. Finally, the role of the State, NGOs and individuals is seen in the larger perspective of facilitating ecotourism. The discussion is based on the presumption that although tourism is an economically profitable activity, it is also environmentally unsustainable. Given the threats to the environment that are a blinding reality today, the paper discusses the move made by the State to accord to the area the status of a National Park. The perceptions of the people with this regard are also highlighted. The paper concludes with a discussion as to whether National Park status can actually ensure sustainable ecotourism in a region and whether it will also result in commensurate economic benefit to the locals. Further, with the restrictions that follow and the entry fee that is levied at each point, will the area be under any less threat than it was before? Can such measures repair the environmental damage that has already occurred? These are some of the issues that the paper raises.

The term ecotourism has been variously defined; and there is no consensus on what it really entails. In 1991, the International Ecotourism Society attempted to define it as "responsible travel to natural areas that conserves the environment and sustains the well-being of local people". Subsequently, in 1996, the International Union for the Conservation of Nature (IUCN now World Conservation Unit), stated that ecotourism is "... environmentally responsible travel and visitation to relatively undisturbed natural areas, in order to enjoy and appreciate nature (and any accompanying cultural features - both past and present) that promotes conservation, has low negative visitor impact, and provides for beneficially active socio economic involvement of local population". It has also been defined as a symbiotic relationship between tourism and the environment which benefits the socio-economic conditions of local inhabitants through the promotion of tourism and the protection of environment. All definitions suggest a confluence of two activities: one that is ecological and the other that involves active participation involving two sets of actors: locals and visitors who constitute the human aspect of ecotourism. It becomes essential therefore, to examine how a society moulds itself according to the needs of the tourists and sustains tourism as an occupation / business. Tourists absorb from the local environment and expect to see an unspoilt tradition bound society that forces many societies to often unwillingly maintain a facade of tradition despite an urge to modernise.

Most competitive tourist industries tend to ignore another vital aspect for the sustenance of the tourist industry anywhere in the world: the environmental aspect. Given the nature of tourism prevalent in most societies, there is often a situation where compromises seem to have been made in favour of the environment. Increasing awareness that money can't always buy everything has led travellers to respect natural environments and cultures. In fragile ecosystems, the question of carrying capacity, that involves how much a region can sustain during a given tourist season without having to stretch its ecologically regenerative capabilities, has assumed great importance. Questions of sustainability have become so vital in present times that the Economic and Social Council of the United Nations has gone as far as to declare the year 2002 as the International Year of Ecotourism. Envisaged as an integral facet of Sustainable Tourism, it is hoped that "sustainable development in the tourist industry will ensure that travel and tourism provide a source of income for many people and contribute to conservation, protection and restoration of the Earth's ecosystem... that environmental protection is an integral part of tourism development" (Resolution 1998/40,46th Plenary Meeting of the Economic and Social Council of the UN). The UNEP sees the promotion of ecotourism as an adjunct to the objective of development[1]. In this paper, an attempt has been made to highlight the elements of competition and conflict (and at times, cooperation) between the two main participants of ecotourism and to assess the possibility of achieving a low impact, minimal, environment friendly tourism given the existence of several interests. With specific reference to the Gangotri region, the paper examines the following aspects of Tourism:

Tourism has provided the people with a means of livelihood and income that has curbed migration to the plains and provided a tradition bound society with alternate means of employment.

It has also transformed the face of tradition and religion. Since the region is part of a popular trekking and mountaineering route in addition to being a pilgrim centre, instances of the Hindu religion being marketed to attract the Western traveller are not uncommon. Infact, heightened tourist activity has led to the establishment of ashrams as micro economic units. It is this growing tendency of religious tourism that the paper examines in addition to discussing its sustainability.

Further, interventions made by the State, various Non Governmental Organisations (NGOs) and individuals in attempting to control environmental degradation and restoring the region to its former glory are discussed. Although these are preliminary, they are nonetheless significant.

An attempt is made rather simplistically, to provide a critique of the current brand of ecotourism, a concept still new to India. Although the avowed claims of ideal ecotourism entails the development of backward regions keeping in view the chief factors of conservation, sustainability and biodiversity, its practice still creates enclaves of beneficiaries leading to unhealthy monopolies. It leads further, to competitive tendencies and tends to work to the detriment of the local population. An analysis is made of the important role played by market forces since each region tries extremely hard to maintain authenticity. In fact, it is interesting to note that the current trend of ecotourism reinforces the very competitive dynamics of market that it seeks to eliminate, thereby leading to more problems of sustainability. The point made in this paper then, is that although ecotourism has led to awareness and preservation of cultures it has not succeeded in minimising the consumerism that commercial tourism generates.

Finally, the paper attempts at resolving the conflicts arising out of the current state of affairs and suggests solutions that if put into practise might suitably alleviate many of the problems that ail the region. This is not to imply that *all* problems will be solved, however, it might be a starting point at arriving at some workable solution that seeks to benefit the local people while at the same time preserving environmental purity.

Using the Gangotri region as an example of highly localised intense tourist activity, this paper tries to bring out elements of co-operation and conflict; and examine attempts at encouraging ecotourism in the area. The observations and conclusions arrived at in this paper are a product of prolonged field research in the region.

The Region

The Gangotri region falls within the Uttarkashi district of Uttaranchal. It is a glacial U-shaped valley which extends eighteen kilometres northwards to the snout of the receding glacier (popularly known as *Gaumukh*[2]). The earliest glacial moraine is to be sighted at Jhala near Harsil and thereafter, there is ample evidence of glacial activity dating to the ice ages. The rock structure comprises largely of limestone and hard granite. The riverbed and the flood plains have a very high content of silt. Towards the snout of the glacier, there is little vegetation. It is estimated that plant growth commences only after several decades have elapsed. However, less than three kilometres beyond, near the region popularly called *Bhojbas*[3] (so called because of the forest of Birch trees in the area), trees and shrubs grow very well. Further south, *Chir* Pine, Maple, Poplar and Juniper form the forest pattern. These forests have been declared Reserved Forests by the Indian Forestry Department. This implies that there are strict regulations and restrictions regarding the use of timber, forest produce and even herbs. The entire region is a unique mix of diverse geographical, demographic and social characteristics, heightened by a highly intensified sense of religiosity. The river Ganga finds worshippers and pilgrims from diverse parts of the country. The mountains which are near the glacier, are revered for being the abode of gods. The region is situated between altitudes of 3000-4000m and is part of the high Himalayas. The Gangotri region is therefore one where every living being, each creature of nature finds itself in the highest order of things and worthy of reverence.

Given the above picture, one would expect that the area to be pristine; the forests preserved and the river unpolluted. However, increased tourism, interest in adventure sports, mountaineering expeditions and pilgrimage, resulted in polluting much of the pristine natural beauty of the Garhwal Himalayas. The Gangotri township has expanded over the years, the number of *ashrams* belonging to hermits has increased and the construction of a direct road to the town has facilitated a larger number of people to access the area. A large area of forest cover around Gangotri, Harsil, Gangnani and Uttarkashi has been cleared for the construction of rest houses, restaurants, shops and so on.

Tourism provides the people with employment (although seasonal) and thus contributes to economic development. Whereas the increase in tourism has controlled large-scale migration of local hill dwellers to the plains in search for jobs. Environmental threats that arise as a result are palpable. These are:

- Deforestation, leading to regular landslides. It is worth noting that deforestation is not always due to timber collection by locals as has often been asserted. The Forest Department has exploited Himalayan forests for commercial purposes; projects such as the construction of roads also necessitate large scale felling of trees.
- The alarmingly rapid recession of the Gangotri glacier which has serious ecological implications. While for the average pilgrim, it heralds the onset of *Kaliyuga*[4]and thus the perversion of humankind; in ecological terms, it demonstrates that the region has been exploited far beyond its regenerative capacity.

Having outlined the major environmental issues in the region, it becomes essential to list the categories of visitors and locals. The visitors who can be classified by the purpose of their travel vary from pilgrims, mountaineers and trekkers. The locals who can be identified by their occupation consist of guides, porters, foresters, priests, restaurant workers and employees in hotels and ashrams. Additionally, local hermits or gurus run ashrams that provide shelter to tourists. These ashrams can also be viewed as micro economic units since they operate on the basis of donation rather than functioning as commercial enterprises and are self-sustaining. It is worth noting that not all people who come to Gangotri in search of employment belong to the neighbouring villages. Many people migrate from Nepal and surrounding areas in search of jobs.

During the season, the State government makes an effort to provide the remote township with adequate Liquefied Petroleum Gas (LPG), Kerosene and even firewood. This has been done to discourage deforestation and to minimise the pollution caused by smoke. There is however, no supply of Hydro Electricity[5] and amenities such as hospitals are absent. The government has also failed to provide basic necessities such as adequate toilets throughout the township. What is even more appalling is that throughout the popular trekking and pilgrim route, there are virtually no toilets and few litter-bins. A few *dhabas*[6] that were permitted by the government, function for the benefit of travellers. However, most supplies take a very long time to reach Gangotri and even longer to be transported beyond. The *dhabas* therefore, run short of stock very often.

Areas of Conflict

Conflicts in the region can be seen at three levels:

Employment vs. Environment: Lured by the prospect of earning a large amount of money locally, locals take up occupations as guides, porters and open hotels and restaurants. While tourism has provided jobs to several people and offered them the luxury of holidays due to its seasonal nature, it has also resulted in felling of trees, the need to construct inns and roads for travellers. In addition, the region has to support an increased population each season. The problem therefore is whether there should be restrictions on the number of people visiting the area since the region cannot sustain large inflows of visitors or, whether there should be controls placed on the locals who must become judicious about utilising natural resources. It is hard to achieve a balance between the two and indeed there is a conflict of interests. With a view to achieving a balance between local's needs and the exigency of environmental preservation, proposals have been made to designate the Gangotri region as a National Park. Since the Gangotri region sustains a unique ecosystem and also a cultural and religious ethos, before classifying the region into a National Park, the needs of the people, the region as providing a major source of livelihood to a large cross section of Garhwali population and the religious significance of the area for most of the country's population should be taken into consideration.

In India, however, the experience of locals with National Parks has been a bitter one. Local populations have been relocated and their access to protected areas that had hitherto provided them with sustenance, severely curtailed. Indigenous populations in the country have developed a symbiotic relationship with their natural habitat and know no other existence. This has made them vulnerable to change and development. In a study conducted in the Nagarhole National Park in Karnataka, where a tiger reserve had displaced tribal populace, it was discovered that they found it hard to adjust to new ways of life. Their numbers had dwindled and they had been forced to eke out an existence from sedentary agriculture. However, their displacement was justified on grounds that they compulsorily hunted for their food and were in direct competition with the tiger for survival (Guha & Martinez-Alier: 1998). Similarly, in the Nanda Devi Bio Reserve, the exclusion of the self sufficient Bhotia tribe from the park has led to a drop in their standard of living and have made them dependent on the State for basic necessities like food (Silori, unpublished). A large part of the problem is that we tend to conform to unrealistic standards of purity; wilderness and naturalness that ignores the human factor as being vital to the environment and instead sees it as destructive.[7] Infact, much of the recent body of knowledge about indigenous plants and herbs comes from the close contact that the people have with their ecosystem. Given this experience, it is hardly surprising that Garhwalis are sceptical about the National Park experiment in the Gangotri region. The preservation of the ecosystem in the Gangotri region involves the sustenance of traditional modes of occupations, a control of the pressure of population on the township during the tourist season so that the pressure over the receding glacier is minimised. Although the logic of a National Park seems desirable, the conflict between environmental requirements and local employment comes through quite clearly, (see Box 1)

Box 1. Inflation and Environment

The economic situation of hill dwellers influences the environment in several ways. Upon a comparison of fuel consumption in Gangotri in the past five years, it was discovered that with time, firewood was preferred to kerosene oil as rationing and the costs of the oil discouraged its use. Also, the plantations nurtured by the GCP that were handed over to the community were found to have been trespassed on many occasions and trees chopped for use as firewood. Costs of procuring LPG are prohibitively expensive and thus this avenue is seldom resorted to. There is also increased evidence of black marketing of kerosene oil. Government raids checking the use of firewood are extremely rare and random. It is important to note that hill dwellers are fully aware of environmental implications. However, faced with a situation where a choice needs to be made between the environment and survival, they really don't have much of an option. When the electric power station at Patangana, two kilometres from Gangotri township becomes operational, firewood consumption is expected to reduce considerably.

Needless to say, the State is heavily criticised for the regulations it imposes regarding the use of forest resources. The Forest Department has been accused of conducting commercial felling of trees, rationing firewood and discouraging locals from collecting the same. A common fear voiced by several locals is that a National Park in the region will hinder local participation and will prevent access to forest produce altogether. Poaching in the form of collection of herbs is also rampant and several people still continue to dodge the forest authorities and collect valuable herbs that are later sold to reputed pharmacies and cosmetic companies.

A check-post at Gangotri has been erected and the trekker or pilgrim to *Gaumukh* and beyond is required to undergo a thorough search and any plastic or non-biodegradable item is promptly confiscated. This can be retrieved on return. A fee of Rs. 50 (approximately $1) is also levied which is returned with the confiscated item. However, the guards at the check post are invariably absent at the time of return of the pilgrim or tourist. Whether this happens on purpose or by sheer coincidence is a matter of debate!

Tradition Vs. *Modernity:* Another dimension concerning the National Park debate involves the opinion of the pilgrims who regard the Gangotri glacier as a National shrine. The creation of a National Park at Gangotri would be tantamount to the imposition of an entry fee on pilgrims and mountaineers visiting the region; thereby curbing the abuse of the environment in an ecologically sensitive zone by restricting the number of people visiting it. However, the idea has not gained much popularity since pilgrims believe that they have a right to freely visit their shrines. The proposal to levy a nominal fee on Indians and a dollar rate on western tourists has found some supporters. Very few believe that there should be parity between all categories of tourists and that they should pay according to the number of days they spend in the National Park. The argument is that any visitor is likely to cause as much ecological damage as the other. However, these are technicalities but they do reflect popular opinion among the people. Interestingly, locals who have more at stake than the visitors have suggested extreme measures such as closing the entire area to all tourists and giving the region sufficient time to regenerate itself and then promoting carefully managed ecotourism.[8]

The above-mentioned conflict between the forces of tradition and modernity manifests itself in another way. This can be seen between the secular tourist industry that hinges on the forces of demand and supply, offers luxury, is client oriented and treats travel as a commodity; and pilgrimage - a deeply religious, low impact, minimalist exercise that produces employment for few and veers on an ethical moral code. For instance, the Gangotri region is considered a vegetarian zone where the consumption of alcohol and meat of any kind is forbidden. Foreign tourists often find it difficult to adapt to these prescriptions. Infact, many pilgrims treat mountaineers and western tourists with contempt since they are often unable to compromise with their dietary habits.

Another point worth mentioning in this discussion is the control of the affairs of the township by the ecclesiastical priestly community. No plans or projects for the development of the area can either be sanctioned or implemented without their approval. The priestly community belongs to the village of Mukhba, 24 kilometres away from Gangotri and was historically given the right to manage the temple township by the Maharaja of Tehri. Since then, this community has jealously guarded any attempt at initiating modernisation and has vociferously protested the projection of the Gangotri region as a tourist spot. This community manages most rest houses and *ashrams* and is the main beneficiary of increased tourism. Surprisingly, this community is responsible for most illegal construction and encroachment all over the township.

Culture Vs. Environment: The environment is equally at risk left to Indian pilgrims, as many pilgrims strip the Indian Paper Birch (*Betula Utilis*), also called *Bhojpatra,* of its bark since it is considered holy. The religious significance of the tree is due to the fact that Hindu scriptures were written on its bark. Even Kalidasa is said to have used it for penning his plays. This contributed to deforestation of Birch trees. Since the time taken for regeneration and growth is much longer at high altitudes, the tree suffers serious environmental threat. Birch plantations in the region are also under threat from ponies that transport pilgrims across to *Gaumukh. Ashrams* run by hermits use firewood for construction and heating since it is fire resistant, cheap and locally available. At *Bhojbas,* the resident hermit continues to send his men far across the hill for firewood since the forest surrounding his *ashram* has already been destroyed. A few tons of wood is burnt each day at his *ashram* despite the fact that he was given five free gas connections. Although *Bhojpatra* is being planted so that the tree is saved from extinction, pilgrims have not been able to resist the temptation of stripping the bark as a souvenir. Efforts at replantation have therefore met with little success. However, with the Birch becoming increasingly rare, pilgrims and saints have resorted to using Juniper for fuel. Many lament that people do not know how to conserve; instead, they are willing to squeeze every little bit from Nature for survival. The pollution caused by local enterprises such as tea stalls and hotels does not need to be stressed. In addition, temple offerings made of polythene cause a lot of ecological damage. Thus far, scant attempt has been made to discourage the use of polythene offerings at the temple in Gangotri or the glacier in *Gaumukh.* This is compounded by the lack of awareness among visitors who are not concerned by the environmental implications of a receding glacier.

Ashrams Promoting Religious Ecotourism

The banks of the river Ganga from the township of Gangotri to Haridwar, where the river is considered to leave her childhood in the hills and head towards maturity in the plains, are dotted with innumerable *ashrams. Ashrams* initially began as retreats where saints lived a life of seclusion and austerity. They were situated on the banks of rivers amid forests and appealed to the pilgrim. As the pilgrim progressed towards the hills, these saints provided shelter and their familiarity with nature was seen as an effective support system. Indeed, these ashrams kept the pilgrim routes alive and also served as guiding points towards the destination. Interestingly, any deviation from the path prescribed by the ashram saints would be tantamount to sacrilege (Singh & Kaur: 1983). A fear of having violated the religious code at disobeying the prescribed route also served the purpose of preservation. The forests and their wealth remained unexplored, access to forest produce was minimal and settlement was limited to areas determined by the saints. With the passage of time, however, they have effectively transformed into semi commercial enterprises that offer shelter and discourses on religion to the pilgrim and the tourist. The saints running these retreats often offer courses in yoga and traditional medicine. For the western tourist, the novelty of staying in austerity has a special charm leading to the popularity of *ashrams* as rest houses. Most *ashrams* also provide shelter free of cost and often settle on a donation depending on the financial capability of the tourist. They provide simple food that is grown on the grounds of the retreat and discourage long-term visitors. Even to this day, they function as transit zones between pilgrim places. With the increase in tourism, there has been a rise in the number of *ashrams* providing cultural retreats leading to a cluster within a limited area. Today, it is not uncommon to find recluse/ hermits encroaching on forest property to build small concrete dwellings for themselves that subsequently often expand into *ashrams* after the surrounding hillside has been blasted. The ecological impacts have also been severe. In Gangotri, demands for fuel have to be met with and not all *ashrams* have gas connections. In *Tapovan,* a high altitude meadow across the Gangotri glacier, where construction of huts and settlement of any kind is prohibited, many saints continue to defy these injunctions. Several *ashrams* have been held responsible for illegal deforestation. Also, the spirit with which *ashrams* were originally established has not been honoured. They have ceased to be transit areas and instead encourage long-term guests who range from the disenchanted western traveller searching for truth to the retired householder seeking detachment. To hold the *ashrams* as directly responsible for the increase in environmental threats would not be entirely true; they however, provide an interesting study as to how ecotourism can still be practised within the confines of culture, provided their growth is controlled. They are also representative of how every section of modem society responds to needs of the hour and moulds itself in a manner so that traditional structures can be sustained (see Box 2).

A change in the nature of *Ashrams* has come about due to the change in the nature of modem day pilgrimages with the improvement of transport and communication. Contemporary scholarship maintains that pilgrim places were characterised by their unique geographical features such as banks of rivers, hot springs, snout of glaciers etc. (Kaur: 1985). A pilgrimage was undertaken for the expiation of sins in order to achieve salvation. The intensity of hardship during the journey implied superior reward and hence simplicity and detachment during the journey was recommended. Viewed in this light, pilgrimages were equivalent to modem day ecotourism since access to these locales was limited due to the hardship associated with the journey. Pilgrimages in the present day are less strenuous and the novelty and achievement associated with the successful completion of a pilgrimage has worn out. The modern pilgrim often makes heavy demands on the environment and does not accord pilgrimage the sacrosanct status that it once occupied. Infact, on the *Gaumukh* trek, a family suffering from altitude sickness and fatigue demanded that the Department of Tourism install a ropeway between Gangotri and *Gaumukh* so that they are not put to the strain of having to trek all the way! Pilgrims now demand luxury and have little time to spend by way of travel.

Box 2. Employers, Exploiters and Wasters

A survey done of ashrams reveals the increasing commercialisation of pilgrimage. Not only have they increased the sphere of their functioning from providing shelter to the devout for a few nights, but have also dispensed with any significant charitable work. They exist in the area as purely commercial organisations and have become extremely mercenary about exacting hefty donations from pilgrims, at times even denying accommodation to those who cannot pay. Most ashrams function as charitable trusts and look to patronage from the wealthy. They provide employment to several people, the largest ashram in Gangotri employs up to 40 people during the peak season. It is interesting to note that few ashrams really employ local people: most of the staff comes from Nepal. None of their employees (except the religious and administrative staff) are permanent, the average monthly salary being Rs. 1500 (approximately $32) inclusive of food. Although not all ashrams have heating facility, an average ashram can bum up to 14 quintals of wood per day. The consumption of kerosene is also significantly high, averaging approximately 1200 litres a month. With the strict rationing of kerosene oil in the region, it is interesting to note that these ashrams still manage to procure generous quantities of the oil. Almost every ashram runs on generators in the evening. One ashram had three generators of varying capacities, the largest of which could provide electricity to the entire township. However, such noble acts of environmental charity are never practised and evenings in the township are abuzz with the drone of several generators. The philosophy behind environmental degradation propounded by the ashram gurus is often misplaced: only one swami interviewed had substantial knowledge of environmental threats in the region. Most gurus prefer to perpetuate the myth of the all-purifying river Ganga and cite helplessness before the almighty as reasons as to why they can only remain instruments in decay rather than prevent the same. During the season, several ashrams initiate sacrificial fires in order to purify the atmosphere. Large quantities of wood is brunt in these ceremonies, the ashram festooned and illuminated with the help of generators and incantations are made for the salvation of souls, asking the almighty to deliver them their trespasses.

Attempts at promoting Ecotourism

Given the intensity of adventure sports and tourist activity in the region, the need to promote ecologically sustainable tourism has been felt by both the State and Non Governmental Organisations. Indeed, it is very difficult to alter the mode of prevalent tourism that is wasteful and exploitative; however, it is not altogether impossible. The State through its agencies is involved in dissemination of information in an attempt to encourage ecologically viable tourism.

The Gangotri Conservation Project (GCP) operates in this region and claims to be the only NGO that aims at greening and cleaning the region. It has installed an incinerator at Gangotri and organised the collection of garbage after the season from *Gaumukh, Tapovan, Bhojbas, Chirbas, Nandanvan* and the *Raktavan* valley. The GCP also started the practice of assigning the task of garbage collection to contractors who were often not committed enough towards the cleanliness drive.

Over the years, the popularity of the incinerator has fluctuated. Once used occasionally, its use soon became regular when contractors were given charge of garbage collection. Its popularity however, soon dipped and it is now sparingly used. As a result, garbage continues to accumulate by the riverside and most hotels and *dhabas* employ their own methods of disposal. The sad fate of the incinerator is due to the fact that neither the GCP nor the municipality wanted to accept responsibility for depositing the garbage for incineration at the required place! The greening programme of the GCP involves planting trees around the area that was once the victim of increasing deforestation. This area stretches from Uttarkashi township up to *Bhojbas.* The GCP maintains a Birch plantation at *Bhojbas,* a nursery at *Chirbas* and a nursery en route to Gangotri. They distribute leaflets about conserving and preserving nature and wild-life. Dustbins were placed and pre-fabricated toilets installed at Gangnani, *Chirbas* (which was washed away in the floods), Gangotri and *Bhojbas* (these were locked and no tourists/ pilgrims could actually avail of that facility). Despite the failure in inculcating toilet habits among pilgrims, the GCP was instrumental in prevailing upon the Forest Department to limit the number of *dhabas* on the trek to Gaumukh and supplying erect tubular frames complete with LPG for the purpose.

In a bid to encourage local enterprise among the people, the GCP has recently encouraged the plantation of walnut trees as a mode of employment generation (See Box 3). This has been highly popular since the people are encouraged to conserve as well as gain income from these trees. The GCP also has an interesting blueprint for the development of ecotourism in the Gangotri region. This involves using Compressed Natural Gas (CNG) vehicles for transporting pilgrims to and from Gangotri, thereby limiting tourist inflow. In order to wield any influence in the region, the GCP has to function through the Forest Department. Thus not only are their activities open to scrutiny, but they also have to deal with bureaucratic tardiness. The GCP comprises people who have retired from public office and can use their influence to draw global attention to the environmental threat to the Gangotri glacier. In this sense, the GCP is an effective lobby even though their activities and results seem sporadic rather than consistent. Over the past six years, the GCP has had to deal with bureaucratic delays, uncooperative locals and lukewarm response from the government and the people. In addition, they are heavily dependent on external sources to fund their projects. Despite all odds, they have continued to diversify and redefine their role in the region. According to Captain Kohli, the head of the Himalayan Environmental Trust (HET), it is time the locals took tourism in their own hands. It has been suggested that the State should stop subsidising mountaineering courses as it has been doing and should convert these institutes into profit making body and earn foreign exchange by offering their courses to westerners. Indians who have already been trained should render their services as guides in order that Ecotourism can take place under the supervision of responsible local people, (see Box 3)

Individual efforts at greening by locals find encouragement and have met with tremendous success. Dr. Harshavanti Bisht, a Reader of Economics at the P.G. Govt. College, Uttarkashi took it upon herself to initiate the plantation of Birch at *a* site between *Bhojbas* and *Gaumukh.* Her nursery at *Chirbas* employs locals and nurtures saplings before they are considered suitable to survive on their own in the plantations. Moreover, her plantations are the product of prolonged research and labour. The mortality rate of her saplings is fairly low and nearly 75% survive. She solicits the participation of the locals and of the *Uttarakhand Vikas Nidhi* that has given her generous financial assistance. Boards requesting tourists and visitors to observe cleanliness and to protect the environment have been put up at intervals along the entire trek to *Gaumukh.* Her plantation has been nurtured after studying the typical forest pattern of the region. The Birch saplings are alternated with Juniper and Poplar, the trees that grow easily on moraine material. Besides, there is also an undergrowth of herbs such as *Rudravanti* and *Thyme.* The presence of these enables the trees to grow naturally, in a forest environment.

The Garhwal Mandal Vikas Nigam (GMVN) is a corporation established by the government to develop in the Garhwal region. It has ample funds and has set up two guesthouses, one at Gangotri and the other at *Bhojbas.* It is required to assist all drives at cleanliness and at greening and cleaning. Many of its employees are actively involved with Dr. Bisht's project. However, these are purely individual inclinations and do not speak much for the organisation which is primarily concerned with making profit during the season. As an organisation, the GMVN needs to become more institutionalised if it is genuinely concerned with regional development. In view of several experts, the GMVN should be privatised and should be concerned with monitoring and promoting tourism rather than indulging in profit based tourist activity.

The initiative taken by the government towards the development of the Gangotri region is extremely crucial.

Box 3. Employment through Environment: The GCP Experiment

After three years of functioning in the Gangotri region, it became apparent to the GCP that motivating the people to conserve the environment unless they were not fully assured of a stable income would be impossible. With a view to encourage the spirit of conservation among the people, aided by the FAO of the United Nations and the Department of Horticulture, Government of India, the GCP initiated the establishment of fast growing walnut plantations that bear fruit in 5 years. 180 walnut saplings were distributed free of cost to the village of Jaspur Taknor, a few kilometres away from Gangotri. 50 families look after these plants and are required to contribute Rs. 5 per month for maintenance. Upon production, profits will be distributed among them. GCP also maintains a nursery with 320 saplings that can be purchased for a nominal rate. Apple saplings are also reared in the nursery. It is hoped that these plantations would encourage the spirit of preservation among the people while at the same time provide them with a stable income. Also, the creation of employment in villages near Gangotri would to a large extent, curb clustering by employment seekers around the township.

The local people are still very heavily dependent on the government as the only agency providing subsidies, food, energy and civic amenities. The government undoubtedly scores higher than the Non Governmental Organisations in providing the people with the basic necessities and welfare in an environment that favours seasonal employment and natural disasters. People always have expectations from the government since relief operations, employment generation are largely conducted by this agency.

In 2001, three new states were created by the Government of India. These were: Uttaranchal, Chhattisgarh and Jharkhand. The hills districts of erstwhile Uttar Pradesh (which included Garhwal) were carved to form the new hill state of Uttaranchal and hopes among the people soared about their future. However, according to recent research in the region, there was greater disillusionment among the people who expected the new state to solve their problems of unemployment and under development overnight. Added infrastructural problems became more glaring and the state government had to face a lot of public embarrassment over issues of employment and education. At the time of writing this paper, state elections have been announced and one can hope that a stable government will insure that the interests of the people are kept in mind while framing policies.

Ecotourism in the Gangotri region is a management issue that can be handled by careful planning. This requires collaborative efforts by the local community, NGOs, the State and visitors; and co-ordination among actors providing services. This planning and monitoring cannot only come from outside agencies: NGOs and the State; but has to be delegated to the people who arguably, are the best judges of their future. However, the managerial aspect of this activity involves the highly specialised nature of ecotourism that has so far systematically excluded locals. Even in cases where there is some local involvement, they are regarded merely as junior partners and learners in the project of implementing ecotourism. Another point worth reiterating is that ecotourism has not succeeded in doing away with commercialisation. It might be a low impact activity in theory, but the costs incurred in maintaining standards, monitoring and creating awareness are extremely high. Ecotourism is a saleable quality and different areas are in competition with one another to attract visitors. Although this ensures that certain standards of purity are maintained, it does not harbour well for those regions that do not subscribe to the competitive spirit generated by the market. Moreover, any ecotourism project demands constant re-evaluation since both the environment and human aspect are changing variables. Erosion and decay is inevitable and standards and norms have to be continually revised and reformulated. These costs should be borne in mind while introducing the practise of ecotourism in third world countries where employment and income generation is uncertain, seasonal and dependent on external factors such as a fluctuating tourist inflow. The current trend is to insist that the visitor to an ecologically threatened ecosystem pay for the visit, no matter how careful he/she is about environmental degradation. Thus, entry to National Parks is hiked so that it becomes prohibitively expensive; subsidised institutions that offer adventure training courses such as hiking, mountaineering, para-gliding and so on, are either forced to shut down or become profit making organisations. The point suggested here is that any increase in cost will not simultaneously repair any ecological damage; infact, it might, on the contrary, discourage environmentally sensitive travellers from visiting the area. The current brand of ecotourism ignores this fact and instead encourages the extravagant (and often wasteful) traveller, one who is prepared to pay for what appears to be 'pristine' and 'authentic'. In fact these are value laden variables and superficial to say the least. In the case of the Garhwal Himalayas where cultural authenticity is the catch phrase, one finds that modernisation has been deliberately delayed so that the 'charm' associated with mountain pilgrim sites is maintained. Ecotourism implies tourism with

Table 1. Visitors to Gangotri and Beyond

Year	Gangotri			Beyond Gangotri		
	Indian	Foreign	Total	Indian	Foreign	Total
1980	61401	-	61401	-	-	-
1985	131018	-	131018	-	-	-
1990	108175	-	108175	2831	-	-
1995	102475	1860	104335	42280	1375	43655
1996	187090	1731	188821	75025	932	75857
1997	258080	1496	259576	134992	1061	136053
1998	214483	1431	215914	110695	970	111665

Source: Department of Tourism, Government of UP

a conscience, something that can hardly be supervised by anyone other than the traveller; freedom and the feeling of being one with Nature; instead, it is the concept of monitoring and guidelines that rob the sense of freedom altogether. The emphasis is not that ecotourism should become such a puritanical activity that it becomes time consuming; it is the paucity of time that has encouraged modem travellers to spend large resources in the pursuit of authenticity.

A problem peculiar to most Third World nations is that they are governed by bureaucracies that cause inordinate delays. Much time is wasted over deliberations and discussions and environmental threats often get sidelined. Most governments are also faced with having to make compromises over issues that involve employment and religion; and are often reluctant to take impopular decisions. The role of elites in marginalizing local people in any developmental policy is also vital. Over the years, local communities have been viewed as being detrimental to development and have been pushed to the fringes and their homelands and the wildlife surrounding the same subjugated and exploited (Gadgil and Guha 1992, Guha 1993). Infact several critics see this elitist bias of excluding indigenous population from their habitat as an extension of the colonial tendency except that now development is used as to justify such displacement. Such an attitude has led planners in India to ignore traditional links between society and nature and has encouraged them to believe that these communities are prime culprits in environmental destruction. This attitude is reflected in all the policies in India that are implemented with minimal sensitivity towards their problem. Given such a scenario, it is vital that there is devolution of power, tourism is decentralised and that a measure of autonomy is given to local bodies that can make judicious decisions regarding their physical environment. To end, it would not be out of place to state that it is possible to encourage local people to determine whether or not tourism at any point needs to be regulated and to monitor tourist inflows in these zones. Strengthening community structures either through the assistance of NGOs or through district and local institutions of the state can be an effective way by which people can be initiated into formal structures and their functioning. Encouraging self-employment through tourism and horticulture will also enable the community to subsequently establish guidelines for practising ecotourism in their region. This will also enable them to benefit by way of profits that have eluded them thus far.

Judging by the discussion, it is clear that awareness among both locals and tourists as regards the threats to the environment is abysmally low. Perhaps the government and NGOs could indulge in an exercise by which they disseminate information regarding the dangers to the environment, the behaviour expected of tourists while visiting the region. Also, since the *ashrams* have a great hold over the people, it would not be out of place to include them in the project of building awareness among the masses. However, what is more difficult to eliminate is the fatalistic attitude that accompanies the explanation behind environmental degradation. Fighting the blind acceptance of affairs as routine and the helplessness that what decays cannot be restored can be a Herculean task and requires a complete change in the mindset.

Finally, the creation and promotion of **satellite areas** would succeed in reducing the pressure on the main temple township. Ecotourism involves dispersing tourist traffic effectively in a manner such that no losses occur to any satellite towns. This could be done by constructing rest houses in these places, by promoting natural beauty, by promoting the handicrafts of the people, their horticultural produce and so on. Additionally, restrictions could be imposed on the entry of vehicles into the township and people discouraged from staying overnight. There could also be a limit to the number of days a pilgrim/ visitor could spend in the region. (Such a stipulation already exists with regard to mountaineers).

However, in implementing all these reforms, the role of the State is extremely crucial as it is in all countries where welfare is the prime concern of the State and NGOs only perform the task of negotiating the spaces that exist between State and society.

Notes

1. It is expected to advance the basic goals of conserving biological (and cultural) diversity, by strengthening protected area management systems (public and private) and increasing the value of sound ecosystems; by promoting the sustainable use of biodiversity, by generating income, jobs and business opportunities in ecotourism and related business networks; and sharing the benefits of ecotourism developments equitably with local communities and indigenous people, by obtaining their informed consent and full participation in planning and management of ecotourism businesses. See also http:// www.unep.org.
2. Literally, cow's mouth.
3. Literally, abode of the *Bhojpatra,* Birch tree
4. The Age of Kali. According to Hindu Mythology, the cosmic world has been divided into four ages, each signified by a further depravation of man. The last of these ages is the Age of Kali, the prelude to destruction of all creation.
5. A hydro power station is currently under construction two kilometres from Gangotri. 250MW in capacity, it has been designed to divert the water from a neighbouring waterfall onto turbines that will then generate electricity. Due to the prolonged winter experienced in the area, it will function for six months each year.
6. Roadside eaters.
7. The guidelines laid by the International Union for the Conservation of Nature (IUCN) influence most decisions involving Natural Areas. A Protected Area by their definition would be *An area of land and/or sea especially dedicated to the protection and maintenance of biological diversity, and of natural and associated cultural resources, and managed through legal or other effective means.* Further, the IUCN has categorised these areas into six types. These are as follows:
   - Category Ia: Strict Nature Reserve: protected area managed mainly for science. The area thus is representative of an ecosystem with peculiar geological and physiological features and species which are scientifically studied or environmentally monitored.
   - Category Ib: Wilderness Area: protected area managed mainly for wilderness protection. The idea behind is to enable a given area to sustain its natural character or influence.
   - Category 2: National Park: protected area managed mainly for ecosystem protection and recreation. This category can be defined as one that encompasses the protection of an area of land or sea for the purpose of ecological integrity of one or more ecosystems for present and future generations. This aims at providing a foundation for spiritual, scientific educational, recreational and visitor opportunities, all of which must be environmentally and culturally compatible. It does not exclude from its purview the existence of indigenous and local people and occupations.
   - Category 3: Natural Monument: protected area managed mainly for conservation of specific natural features.
   - Category 4: Habitat/Species Management Area: protected area managed mainly for conservation through management intervention. This area involves the management of a natural area in order to ensure the maintenance of habitats and to meet the requirements of specific species.
   - Protected Landscape/Seascape: This involves the protection of an area largely for the purpose of conservation and recreation. This area may include one where the interaction of people with nature over time has produced an area of distinct character with significant aesthetic, ecological and /or cultural value, and often with high biological diversity. Safeguarding the integrity of this traditional interaction is vital to the protection, management and evolution of such an area.
   - Managed Resource Protected Area: protected area managed mainly for the sustainable use of natural ecosystems. This area consists largely of unmodified natural systems, managed to ensure long term protection and maintenance of biological diversity, while providing at the same time a sustainable flow of natural products and services to meet community needs.

   (Source: http://www.wcmc.org.uk/protected_areas/data/define.htm)
8. At the time of writing this paper, it was discovered that the Gangotri region had already been declared a National Park. When a map demarcating the area covered by the Park was produced, it became evident that the region frequented by pilgrims did not fall within its purview!

References

Gadgil, M. and Ramachandra Guha 1995. Ecology and Equity: *The Use and Abuse of Nature in Contemporary India.* New Delhi (Reprint), Penguin Books India.

Guha, R. and Juan Martinez-Alier 1998. Varieties of Environmentalism. New Delhi, OUP.

Kaur, J. 1985. *Himalayan Pilgrimages and the Neto Tourism.* New Delhi, Himalayan Books.

Silori, Chandra Shekhar, Socio Economic and Ecological Consequences of the Ban on Adventure Tourism in Nanda Devi Biosphere Reserve, Western Himalaya. Unpublished, communicated for the proceedings of the National Seminar on Wild life Conservation Research and Management held at Wildlife Institute of India, Dehradun, Aug 1998.

Singh, T.V. and Jagdish Kaur (eds.) 1983. *Himalayas, Mountain and Men.* Lucknow, Print House India.

www.unep.org

www.wcmc.org.uk/protected_areas/data/define.htm

Globalization, ecotourism and the promise of development in sub-Saharan Africa

Andrew Lepp

This paper examines the current phenomenon of rapidly increasing ecotourism in parte of Sub-Saharan Africa, especially regarding ecotourism's impacts on the people of the region. The discussion takes place within the context of globalization as tourism is a tangible manifestation of increasing global interconnectedness. Within this context, ecotourism can be characterized as a force actively promulgating Western environmental ideologies around the globe. The paper argues that these environmental ideologies, which originate within the world's most technologically developed nations, may not be appropriate for a region of the world largely inhabited by subsistence farmers. The crux of the matter then becomes, how to realize ecotourism's promise of development while minimizing its ideological intrusiveness. The paper concludes by suggesting an approach for tourism and development which is more appropriate for the circumstances of Sub-Saharan Africa.

Now that the new millennium has arrived, an accurate depiction of this long anticipated juncture in human history is finally possible. Among the many reasonable interpretations of these times, is the notion that this is an era of increasing global interconnectedness. Globalization is a tern that has Ibeen used to refer to this very phenomenon (McGrew 1996). Media and computer technology are but one agent of tins in end as they provide instant news and information to any corner of the globe. Similarly, advances in transportation networks contribute by providing access to destinations around the world with hardly more than a day's journey. A tangible manifestation of this increase in global accessibility and interconnectedness is the tourism industry. Like globalization, tourism depends on the movement of information and people around the globe and also is expected to increase in the new millennium. The World Tourism Organization forecasts that international tourist arrivals will climb to 1.6 billion by the year 2020 (UNESCO 1999). No where is this trend of increasing tourism more evident than on the African continent. Worldwide, international tourist arrivals from 1993 to 1997 increased 17.7 per cent (WTO 1999). During the same time period, tourist arrivals in Africa increased 24.9 per cent. Even more amazing, tourist arrivals in Sub-Sahharan Africa increased 56.3 per cent during tins period. Tins rapid growth in Sub-Saharan Africa's tourism industry is one indication of globalization's increasing presence in the region. For the researcher, this phenomenon raises two important questions. First, what are the reasons behind tins increase in tourism to Sub-Saharan Africa; and second, what are some possible consequences of this trend for the people of the region?

The purpose of this paper, however, is not simply to answer the above two questions. instead, these questions will guide a discussion that examines tins recent touristic phenomenon in Sub-Saharan Africa within the context of globalizaion and development. The discussion is grounded in relevant tourism and globalization theory. in addition, the discussion incorporates insights gained from the author's two years of experience working on tourism development projects in Uganda, as well as information gathered from interviews with influential Ugandan tourism officials. The discussion indicates that although tourism has potential as a development tool in Sub-Saharan Africa, it often contributes to globalization's increasingly intrusive flow of Western ideologies throughout the region. It is argued that these Western ideologies may not be appropriate for a region of the world largely inhabited by peasant farmers. The paper concludes by suggesting an approach for tourism and development more appropriate for the circumstances of Sub-Saharan Africa.

Tourism and the African Environment

In the hustle and bustle of modem life in the Western world, daily pleasure is often foregone in favour of capital accumulation. Free time is continually traded for money (Schor 1992). When a person finally becomes too exhausted to continue working, a portion of the strenuously earned capital may be spent on pleasure. The tourism industry offers that pleasure. Crompton (1979) provided this insight when he postulated that the monotony of daily routines in modem life creates a state of disequilibrium within the individual. He then suggested that one way to restore equilibrium is by traveling to a novel destination and breaking the monotonous routine. It seems this has become one of tourism's roles in modem society.

Similarly, Dean MacCannell (1973) has theorized that the modem tourist, living a life of constant stress focused on accumulating capital, has lost touch with a deeper meaning for life. Reconnecting with this meaning is the incentive of modem tourism. As MacCannell explains, "the concern of modems for the shallowness of their lives and inauthenticity of their experiences parallels concerns for the sacred in primitive society" (p. 590). In this sense, tourism is the search for something real, something authentic. From the perspective of a modem tourist, what is real and what is authentic is that which has not yet been developed or modernized, or to use the terminology of MacCannell, still found in a "primitive" state.

MacCannell's use of the words "modern" and "primitive" indicate that the concept of time is crucial in defining the relationship between tourists and hosts. This unequal temporal relationship places the host in a time before the present. In the case of Africa, the reference point is still the continent's imperial past (Behdad 1994). In this sense, it categorizes the host as something "other" than the tourist and therefore as a spectacle itself. Neumann (1997) argues that Africa today is still constituted as "primitive" in the West. Adding to the previous discussion, he argues that the word itself has more than temporal connotations. There is also a spatial dimension to the construction of "primitive." This is especially important for tourism because it means there are places where the primitive can still be found and visited.

Tourism's search for the primitive in Africa dates to the late eighteenth century and became popularized with the publication of travelogues of the early journeys to the continent(Pratt1992). These travelogues remained popular throughout the nineteenth century. By the early twentieth century, the travel exploits of U.S. President Theodore Roosevelt added to the continent's image as a wild and primitive destination. On safari in Africa from 1909 to 1910, President Roosevelt, along with over two hundred African porters and trackers, shot over 3000 specimen of wildlife (Akama 1996). The subsequent popularity of Tarzan novels and movies further contributed to Africa's image as primitive (Keim 1999). This "othering" of Africa continues today with the influence of popular media like *National Geographic* magazine, television and film (Lutz and Collins 1993, Keim 1999). Thus, in reply to this paperi s first question: what are the reasons behind this increase in tourism to Sub-Saharan Africa? It is suggested that the West's construction of Africa as "primitive," and the Western tourist's search for the authentic are two contributing factors.

Ecotourism and Globalization in Africa

MacCannell's (1973) suggestion that tourism is the search for something authentic and not yet developed or modernized also serves as a worthy explanation for the general trend, seen throughout the world, of an increased interest in nature based tourism. In contrast to the modem city, what could be more real or authentic than nature? A recent study found that while tourism as a whole is growing at approximately 4 per cent, nature based tourism is increasing at annual rates between 10 and 30 per cent (Reingold 1993). This trend bodes well for Sub-Saharan .Africa's tourism industry where national parks and wildlife reserves form the major attractions (Travel and Tourism Intelligence 1996). As a nature based destination, it is likely that a significant number of the tourists to Sub-Saharan Africa are Ecotourists. According to 'The *Ecotourism Society* (1998), ecotourism should be considered a segment of the larger nature tourism market. Ecotourism is a relatively new phenomenon in tourism and agreement on a suitable definition has been a matter of debate, yet a congruence of themes has emerged (The Ecotourism Society 1991, Buckley 1994, Ceballos-Lascurain 1996, Blarney 1997, Weaver 1998). The central themes of ecotourism are: travel to natural areas that is low impact, sustainable, and fosters environmental awareness in the tourist; travel that provides direct financial benefits for conservation; and travel that sustains the wellbeing of local people. By meeting these criteria, ecotourism strives to distinguish itself from mass tourism, an industry with profit maximization as the guiding principle.

For this reason, ecotourism has been widely embraced by many African governments as a rural development strategy (Brown 1998). Until recently, development has been defined by certain economic variables such as Gross National Product or per capita income. However, such definitions ignore important variables such as the distribution of wealth, education and equality. Ideally, development should be defined as a process of positive change that occurs over time and results in improved standards of living for the general population. In other words, development should consider not only measures of wealth, but also education, health, gender equality, and access to some form of democratic process. Some of the countries most actively pursuing ecotourism as a development strategy are South Africa, Kenya, Tanzania, Zimbabwe, Uganda, Namibia, Botswana, Zambia, Madagascar, Niger and Cameroon (Weaver, 1998).

The Flow of Western Ideologies

According to the World Conservation Union's (IUCN) publication, *Tourism, ecotourism, and protected areas* (Ceballos-Lascurain 1996), ecotourism has its roots in the environmental movement which originated in the western industrialized nations during the late 1960s. Conservationists at the time began lobbying governments to set aside land not just for tourism purposes but also for the preservation of entire ecosystems. The whale watching industry developed at this time and demonstrated ecotourism's potential for success at generating revenue and promoting conservation. Reflecting on this, the same IUCN publication hails ecotourism as "the new environmental paradigm" (p. 26). Certainly, the foundations of ecotourism are tied to western environmental ideologies. Considering the increase of ecotourism throughout Sub-Saharan Africa, a flag of caution should be raised regarding the phenomenon's Western-style conservation underpinnings. If ecotourism is to separate itself from mass tourism and provide for the well being of people in Africa, one must ask if its Western-style nature conservation ethics are best suited for the job. This is based on the observation of Olwig and Olwig (1979) who surmised that the perception and use of nature held by peasants must be quite different from that of the world's most advanced capitalist societies. Such a flag was recently raised by Del Gizzo (1998) who suggests ecotourism be viewed as the ecological branch of globalization. As such, characterizing ecotourism as a distinct form of tourism becomes more difficult. This is evident when analyzing the interplay between tourism and globalization.

Tourism is the flow of people in search of pleasure. Globalization has also been described in terms of flow (Appadurai 1996). Appadurai identifies five dimensions of global flows which he labels: ethnoscapes (the flow of people); mediascapes (the distribution of channels of information and the images which flow through them); technoscapes (the flow of technology); financescapes (the flow of capital); and ideascapes (the flow of political ideologies, often of an enlightenment or Western world view). Each of these flows can be seen occurring within tourism (Macleod 1999). Tourism involves moving people across national borders. Tourism is supported by a global media campaign ripe with images from around the world, images which can contribute to the "othering" of non-western people (Behdad 1994; Akama 1996; Keim 1999). Tourism relies on vast technological networks as is seen in flight and hotel reservation systems. Tourism creates financial markets and moves capital across international borders. And finally, as this paper strives to demonstrate, tourism facilitates the transfer of ideologies among culturally different people. Clearly, global flows are embedded in tourism. As such, tourism is increasing the interconnectedness of the world. Likewise, so is globalization (McGrew 1996).

Hall (1996) refers to globalization as an uneven process in which individuals are related to global flows in very distinct ways. This relationship is a function of capitalistic expansion (Kothari 1997). Therefore, the people with the greatest access to global flows are those with the most capital, or those of the Western world. This holds true for international tourism, which is an expensive leisure pursuit. Not surprisingly, nearly all international tourists originate in the western world. In light of this, tourism's flows are not equal. The direction of these flows is primarily from the developed world to the underdeveloped world. This parallels the larger trend of globalization and is yet another example of how tourism and globalization are inherently linked.

As Appadurai (1996) suggests, this flow of tourists brings with it a constant flow of ideas. One of the newest ideas circulated by tourism's flow is conservation of the natural world. This is seen in the increase of nature tourism (Reingold 1993). This ideology of conservation is embedded in ecotourism (Del Gizzo 1998). In this regard, Del Gizzo provides a more practical spin on the ecotourism phenomenon. Ecotourism uses the promise of rural development to persuade underdeveloped countries to conserve their natural environment with Western-style tactics. Proof of this is evident in the West's tremendous funding of ecotourism related development projects. For example, by the mid 1990s international aid agencies were providing billions of US dollars for development projects with ecotourism related components, including more than $2 billion in funding from the U.S. Agency for International Development (USAID) (Honey 1999). Ecotourism's potential to generate revenue is made possible by Western tourists in search of the authentic (MacCannell 1973) who romanticize ecotourism attractions in developing countries as the last remaining vestiges of a wild and primitive earth. In this way, ecotourism purveys the flow of Western conservation ethics and ideologies as well as stereotypes of "the other."

This is clearly the case in Sub-Saharan Africa where the vast majority of inter-regional tourists are from industrialized, Western nations, and the natural environment is the major attraction (Travel and Tourism Intelligence 1996). The result is that ecotourism is largely influenced by Western environmental ideologies. Unfortunately, these Western ideologies are not necessarily those of the African people. Accordingly, this is an issue where ecotourism has recently been criticized. Guha (1989) accuses international conservation agencies of promoting an American style, preservation oriented park system throughout the developing world. While equating conservation with wilderness preservation has worked in the unique context of the US, it is not appropriate for most of the world, in fact, Guha calls it "positively harmful" (p. 75). Guha claims these parks are managed for Western tourists and pit the interests of wildlife against those of rural people neighbouring the park. The result, Guha claims is a transfer of resources, in the form of protected parks, from the poor to the rich - or in the case of ecotourism, from local people to foreign tourists.

This can be seen in an African context where rural people of Kenya have not rushed to support the Western-style conservation values ecotourism development because the targeted resources are needed for meeting their subsistence needs (Akama 1996). Furthermore, in Africa, wildlife is often viewed as the enemy of the people. In the Tunduru district of Tanzania bordering the Selous Game Reserve, forty-eight people were killed and thirty-six injured due to lions between 1985 and 1988, this is in addition to other deaths caused by crocodiles and rampaging elephants (Songorwa 1999). Not surprisingly, a study in this district found that local people were not interested in conserving wildlife on their land (Songorwa 1999). In fact, the study revealed that the people in this part of Tanzania were more interested in having legal access to natural resources than conserving them.

Thus, in contrast to the growing Western view of the environment as a sanctuary to be preserved, African values hold the environment to be a provider (Mugisha pers. comm.). The natural environment sustains life by providing food and water, building materials, and medicines. Today, as African populations steadily increase, the need to access these natural resources is becoming more and more important. In contrast to this need, ecotourism, by creating a demand for "primitive" environments, renders these natural resources inaccessible to everyone but tourists. The unspoken idea is to "freeze" a tourist attraction in its mythic, authentic state (Del Gizzo 1998). The result is the disempowering of local people as they lose access to important resources. Surely this runs contrary to ecotourism's stated purpose of sustaining the well being of local people.

Thus, Western environmentalism and ecotourism have worked to alienate Africans from natural resources. For example, a now widely quoted study conducted in Amboseli National Park determined the value of one maned lion to be $515,000 in terms of tourist generated foreign exchange (Thresher 1981). The study then computed that the value of one maned lion was comparable to 30,000 head of Maasai cattle. The underlying argument was that lions are much more valuable than Maasai cattle and that Maasai cattle should be removed from the park and surrounding ecosystem in favor of tourism. However, what the study fails to consider is the value that the Maasai place on their cattle, and on their cultural tradition of cattle keeping. In fact, the Maasai need cattle as bridewealth to marry women. Cattle are a symbol of status. Cattle serve as the Maasai's bank, keeping their money safe from inflation, and cattle are the source of daily food. A common East African adage explains "the bone of the human being rests on the bone of the cow" (Mugisha 2000 pers. comm.). Clearly, this value can not adequately be measured in terms of dollars. Thus, in some cases, the values and ideologies of Africans contrast dramatically with those held by Western conservationists and ecotourists. The resulting conflict seems to be one reasonable consequence of this trend of increasing tourism in Sub-Saharan Africa.

The Flaw of Capital and the Decreasing Role of the African State

As the above discussion shows, Africans living around national parks are dependent on the same resources as the multi-billion dollar tourism industry. In the struggle for these resources, it is often the powerful tourism industry that is cutting the fatter hog. Foreign capital now has a major presence in every sector of the African tourism industry. The West's insistence on neo-liberal trade policies towards Africa, such as widespread structural readjustment programmes, has opened borders for multinational corporations including those of the tourism industry (Nyang'oro and Shaw 1998). This has allowed the penetration of multinational corporations who easily out- compete local tourism enterprises. In Africa, 72% of hotel investment is linked to multinational corporations. For comparison, only 2% of hotel investment in Western Europe is linked to multinational corporations (Honey 1999). Foreign investment in tourism means that capital generated by the investment will leave the country and return to the investor. It has been estimated that 55 per cent of tourism dollars leak out of developing countries via foreign owned tour operators, airlines, hotels and by locally owned operators paying for imported food, drinks, and other specialty items. Some studies estimate this leakage to be 80 to 90 per cent (Brown 1998, Honey 1999).

This vast flow of capital out of Africa should not be surprising considering tourism's symmetry with globalization. Appadurai (1996), in constructing the five dimensions of globalization, recognized that the flow of capital is an important factor. That capital tends to flow out of Africa to the West is a classic symptom of dependency. Dependency theory of underdevelopment (Frank 1967) holds that the linkage of underdeveloped countries with developed, industrialized countries is a malignant relationship for the former and a beneficial one for the latter. The effect of this linkage is decapitalization (or the extraction of capital) that frustrates the development potential of third world countries. The current economic reforms encouraged throughout Africa by Western agencies encourage this capital flight (Nyang'oro and Shaw 1998). This phenomenon also supports Hall's (1996) contention that globalization, including the flow of capital, is an uneven process. This is contrary to the focus of accepted development strategies, which strive to raise the living standards for the majority across all economic sectors and geographic regions (Brohman 1996). Instead, globalization's flow of capital is a process centered in the West (Shaw 1997).

A repercussion in Africa, besides widespread poverty, of structural readjustment is the decline of the role of the African state (Nyang'oro and Shaw 1998). Power is shifting to the multinational corporations, lending agencies, and NGOs that are penetrating the region. This is also reflected in the decision making process surrounding ecotourism developments on the continent. For example, in Uganda, it is estimated that 80 per cent of the money used to manage the country's national parks is provided by foreigners (Mugisha 1999 pers. comm.). These parks represent the country's primary tourist attractions. The result of this severe dependency on foreign money is that the Uganda Wildlife Authority (UWA) has lost its autonomy to manage its own parks and natural resources. In 1998, the World Bank was UWA's primary source of funding. This funding came with the stipulations that UWA's financial operations be managed by a private firm, that UWA recruit a new Executive Director from outside of Uganda, and that various other UWA operations such as some tourist concessions be privatized (Mugisha 1999 pers. comm.). Thus, accepting donor funds translated into a clear loss of power for UWA, which is best represented by UWA's hiring of a foreign Executive Director. Uganda is not a special case. In fact, most African governments' conservation strategies are influenced by the priorities of external donors (Derman 1995).

A major problem with this trend is that the international NGOs and other agencies providing funds and making decisions in Africa do not have to answer to the African people. That is the responsibility of the state. Instead, these international agencies have a constituency far removed from the African continent which sets their agenda and to whom they must answer (Derman 1995). For example, in Uganda, does the Executive Director of UWA answer to the World Bank or to the Ugandan people whose resources he is charged to manage? The answer seems clear and the effect of this is further disempowerment of African people as they are removed from any political process. As Alcorn (1993) succinctly states, "conservationists are acting as gatekeepers to a discussion table that does not have a place set for those whose homeland's future hangs in the balance" (p. 426).

Ecotourism, Democracy and Africa

This discussion has tried to illustrate how ecotourism in Africa, operating as an arm of globalization, perpetuates an uneven power structure, centered in the West, which excludes local people. The values and ideologies of local people are ignored in favor of Western values, natural resources are being managed for the pleasure of Western tourists, ecotourism revenues often leak out of the country to Western financial centers, and finally, local people are left out of decision making processes as power has shifted to Western agencies with no responsibility to African people. Therefore, ecotourism's stated goal of sustaining the well-being of local people is being seriously obstructed and it may be failing as a development strategy. Ideally, any development strategy should involve a democratic process. When people lose the power to decide how their shared resources are best used, democracy is thwarted. If democracy is thwarted, development is thwarted. Could this in fact be a second consequence of increasing tourism to Sub-Saharan Africa? If so, it runs contrary to all the promises made by the proponents of ecotourism in the international donor community. In the face of a tourism industry predicted to quadruple by 2020, this is a troubling scenario, and one that forces the investigation of another question: is this the ultimate result of ecotourism in Africa or can a more empowering, democratic form of ecotourism be created?

To answer this question, it is helpful to once again think of tourism within the larger context of globalization. Claude Ake (1995) has theorized that the consequence of globalization is the reconstruction of the world in terms of economic forces. In this new world order, a nation's position in world governance corresponds to that nation's economic power, the global order is presided over by an informal cabinet of economic elites, and democracy succumbs to the logic of the market. It is this insight which forces Ake to recognize that at this moment "there is no part of the world where democracy comes remotely close to being taken seriously" (p.30). Even in the US, the sovereignty of the people is under relentless pressure from the rise of sovereign capital. Ake notes that in this process, "government by the people" is being replaced with "government by consent of the people" (p.37). This is marked by the increasing triviality of popular participation as was seen in the 2000 US presidential election where the final result was decided by the politically appointed supreme court. 'The end result, as Ake notes, is that the right of property prevails over the will of the majority. In the global arena, this is effectively disempowering the world's poorest nations. An in depth illustration of this has already been provided in the above discussion of tourism.

The only solution is that democracy be taken seriously again, beginning with a return to its roots - people. In the African context, this means revisiting the rural areas. The majority of people in Sub-Saharan Africa live in the countryside and are not fully connected with national political structures (Owusu 1992). Therefore, Owusu suggests new institutions and structures of democracy should be adapted to rural traditions and concepts of community. It is now that the discussion can return to ecotourism, an activity which in Africa, is based in rural settings around wildlife parks and preserves. For ecotourism to fulfill its promise of economic opportunity, it must empower local people through democratic processes based on local participation and consensus in the tourism project. No tourism development should claim to be ecotourism unless it meets this most important of all criteria.

A brief explanation of the political economy of ecotourism development in East Africa illustrates the breadth of this challenge. The East African agencies responsible for park management each have a special branch with the responsibility of promoting community development (Mugisha 2000, Honey 1999). Ecotourism is one obvious method of achieving this objective. Therefore, ecotourism development in Kenya, Tanzania and Uganda is largely facilitated by government agencies as a part of larger park management projects (Honey 1999). In practice, this amounts to a top down approach because ecotourism development strategy originates with the central government, or the Western agencies funding the central government's operations. This has excluded the local people from the genesis of many centrally directed projects. As a result, the projects themselves are perceived to be serving government interests instead of the local people's. As IUCN's former Uganda Programme Coordinator expressed, "given current government policies, I do not expect very much grass roots tourism ... most of the tourism industry is under the control of the central government, local communities can only benefit indirectly; at the same time, the potential for grass roots tourism is very high" (Sebunya 2000).

Despite current practices, increased democracy in ecotourism development is the desire of African people. As the former Ugandan country director for IUCN recently commented, "Community involvement has to increase given our political structure. Political democracy and transparency will increase and local officials will be elected on what they can do for the communities, those near parks have already started asking for more involvement into park management" (Sebunyo 2000 pers. comm.). Likewise, the Uganda Wildlife Authority's Tourism Service Coordinator stated "increased community participation in tourism" as a major trend for the years ahead (Ajarova 2000 pers. comm.). Thus, for ecotourism to keep its promise of development and to distinguish itself as separate from mass tourism, it must include democratic process as one of its primary tenets.

Conclusion

Tourism is not alone in its classification of Africa as the "other." Yes, the construction of Africa as a primitive destination couples well with MacCanneli's (1973) classic explanation of tourism as a search for the authentic. However, the "othering" of culeras of non-European origin has been a fundamental element of European identity formation since the beginning the imperial era (Pratt 1992). The task now is to emancipate ourselves from this practice. Ecotourism can be a mechanism for achieving this goal if it is re-conceptualized to accommodate this concern. This demands more from ecotourism than simply showing cultural sensitivity. In addition, it requires the incorporation of the host culture's values, goals and needs into the design of the ecotourism project. This holistic approach is necessary to counter the prevailing dominance of Western ideologies.

The matter at stake is how to realize ecotourism's promise of development along with all of the accompanying tenets such as education, health and equality, while the intrusiveness of Western ideologies. The solution is clearly to incorporate fundamental democratic principles into the process of ecotourism development. Not the illusion of democracy prevalent in the West where capital is sovereign over the individual, but democracy in a rural African sense. In other words, rural African values should be at the center of any ecotourism development process. Thus, ecotourism will only make good on its promises if it can progressively remodel itself to be more in time with the traditions, beliefs, and values of the rural people it proposes to help. In this way, ecotourism can be an agent of consensus, participation, tolerance, and equality, as well as the fair distribution of the burdens and rewards of conservation. However, without a rural African sense of democracy as a guiding principle, ecotourism will remain fettered to globalization and Western environmental ideologies. It will therefore continue to compete with African people for valuable resources, exclude them from greater economic opportunities and leave its promise of development unfulfilled.

References

Ajarova, L. 2000. Tourism Development Manager, Uganda Wildlife Authority. Personal communication.

Akama, J.S. 1996. Western Environmental Values and Nature Based Tourism in Kenya. *Tourism Management*, 17 (8): 567-574.

Ake, C. 1995. The New World Order: a view from Africa. In H.H. Holm and G. Sorensen (Eds.) *Whose World Order? Uneven Globalization at the End of the Cold War.* Boulder, Colorado, Westview Press.

Alcorn, J.B. 1993. Indigenous Peoples and Conservation. *Conservation Biology*, 7(2): 424-426.

Appadurai, A. 1996. *Modernity at Large — Cultural Dimensions of Globalization.* Minneapolis, MN, University of Minnesota Press.

Behdad, A. 1994. *Belated Travelers: Orientalism in the Age of Colonial Dissolution.* Durham, Duke University Press.

Blamey, R. 1997. Ecotourism: the Search for an Operational Definition. *Journal of Sustainable Tourism*, 5(2): 109-130.

Brohman, J. 1996. New Directions for Tourism in Third World Development. *Annals of Tourism Research*, 23(1): 48-70.

Brown, D.0.1998. In Search of an Appropriate Form of Tourism for Africa: Lessons from the Past. *Tourism Management*, 19(3): 237-245.

Buckley, R. 1994. A Framework for Ecotourism. *Annals of Tourism Research*, 21(3): 661-669.

Burns, P. 1999. Paradoxes in planning: tourism elitism or brutalism? *Annals of Tourism Research*, 26(2): 329-348.

Ceballos-Lascurain, H. 1996. *Tourism, Ecotourism and Protected Areas.* Switzerland, IUCN.

Crompton, J. 1979. Motivations for Pleasure Vacation. *Annals of Tourism Research*, 6:408-424.

Derman, B. 1995. Environmental NGOs, Dispossession, and the State: The ideology and praxis of African Nature and Development. *Human Ecology*, 23(2): 199-215.

Del Gizzo, S. 1998. Peephole to the Jungle: A study of safari from imperial preservation to global preservation. *Proteus*, 15(2): 37-42.

Frank, A.G. 1967. *Capitalism and Underdevelopment in Latin America.* New York, Monthly Review Press.

Guha, R. 1989. Radical American Environmentalism and Wilderness Preservation: A Third World Critique. *Environmental Ethics*, 11: 71-83.

Hall, S. 1996. The Question of Cultural Identity. In S. Hall, D. Held, D. Hubert and K. Thompson (Eds.) *Modernity.* Oxford, UK, Blackwell Publishers.

Honey, M. 1999. *Ecotourism and Sustainable Development. Who Owns Paradise?* Washington, D.C., Island Press.

Keim, C1999. *Mistaking Africa: Curiosities and Inventions of the American Mind.* Boulder, Colorado, Westview Press.

Kothari, R. 1997. Globalization: A world adrift. *Alternatives*, 22: 227-267.

Lutz, C.A. and J.L. COLLINS 1993. *Reading National Geographic.* Chicago, University of Chicago Press.

Maccannell, D. 1973. Staged Authenticity: Arrangements of Social Space in Tourist Settings. *American Journal of Sociology, 79:* 589-603.

Mackenzie, J. 1988. *The Empire of Nature: Hunting, Conservation and British Imperialism.* Manchester, UK, Manchester University Press.

Macleod, D.V.L. 1999. Tourism and the Globalization of a Canary Island. *Journal of the Royal Anthropological Institute*, 5(3): 443-457.

Mcgrew, A. 1996. A Global Society? In S. Hall, D. Held, D. Hubert and K. Thompson (Eds.) *Modernity.* London, Blackwell.

Mugisha, A. 1999. Director of Field Operations, Uganda Wildlife Authority. Personal communication.

Mugisha, A. 2000. Director of Field Operations, Uganda Wildlife Authority. Personal communication.

Neumann, R.P. 1997. Primitive Ideas: Protected Area Buffer Zones and Politics of Land in Africa. *Development and Change* 28: 559-582.

Nyang'oro, J.E. and T.M. SHAW 1998. The African State in the Global Economic Context. In L.A. Villaton and P.A. Huxtable (Eds.) *The African State at a Critical Juncture.* Boulder, CO, Lynne Rienner.

Olwig, K.F. and K. Olwig 1979. Underdevelopment and the Development of "Natural" Park Ideology. *Antipode*, 11:16-25.

Owusu, M. 1992. Democracy and Africa - a view from the village. *The Journal of Modem African Studies*, 30(3): 369-396.

Pratt, M.L. 1992. *Imperial Eyes: Travel Writing and Transculturation.* New York, Routledge.

Reingold, L. 1993. Identifying the illusive ecotourist. *Going Green: A Supplement to Tour & Travel News*, October 25: 36-37.

Schor, J. 1992. *The Overworked American.* New York, BasicBooks.

Sebunyo, K. 2000. Former Uganda Country Director, IUCN. Personal communication.

Shaw, M. 1997. The State of Globalization: towards a theory of state transformation. *Review of International Political Economy*, 4(3): 497-513.

Songorwa, A.N. 1999. Community-Based Wildlife Management (CWM) in Tanzania: Are the Communities Interested? *World Development*, 27(12): 2061-2079.

The ecotourism society 1998. Ecotourism Statistical Fact Sheet, www.ecotourism.org

Thresher, P. 1981. The economics of a lion. *UNASYLVA* 33(134): 34-35.

Travel and tourism intelligence 1996. Europe to Sub-Saharan Africa. *Travel and Tourism Analyst*, 5: 46-64.

Unesco 1999. *UNESCO Courier*, 52(7): 26.

Weaver, D.B. 1998. *Ecotourism in the* less *Developed World.* New York, CAB International

World tourism organization 1999. *Yearbook of Tourism Statistics.* Madrid, World Tourism Organization.

Local initiatives to localize ecotourism: An exploratory study in Kerwa, Van Vihar National Park catchments

A. K. Bhattacharya, Suchitra Banerjee and Vandana Saksena

Sustainable ecotourism management and planning requires the prudent support of local population of the destination area. The following study attempts to explore a few aspects of community based ecotourism of an essentially local area, in the close proximity of Bhopal, the capital of Madhya Pradesh, India. A preliminary study was initiated in the Kerwa forest area, in an attempt to profile the local villagers in order to identify common concerns. Personal visits, formal and informal communication, and discussions were employed to gather primary information to test the feasibility of initiating any attempt towards the sustainable development of ecotourism. Secondary data on the aspects of demand and supply was largely verified and utilized to profile the visitors and the visited. The data thus collected was synthesized and analysed to arrive at conclusions and strategies in the required direction.

From Tourism to Ecotourism

Despite abating forces of economic recession, terrorism and political instability, tourism is reckoned as one of the most important civil industries in the world. Within the vast gamut of the phenomenon, ecotourism holds the biggest promise for the developing countries of the world (WTO 1984, 1991). Spurred by the Brundtland report, followed by the Rio summit and the recent Kyoto Protocol, environmental consciousness has taken precedence in all human endeavours. Eco-development became the catch phrase. It aimed at integrating cultural, social, and ecological goals with development (Sagasti and Colby 1993; Nelson 1994). Tourism, as an industry, responded in tandem.

The concept of ecotourism evolved in 1980s from the commercialized chaos of unsustainable forms of tourism (Figgis 1993; Wight 1993; Figgis 1994; Harris and Leiper 1995; Eagles 1996). Embracing ecotourism as a means of 'giving nature value' (Kamra 2001), almost took away the sting from fordian tourism forms. When the term ecotourism first appeared in the 1980s, it was loosely applied to travel that included any aspect of nature. Butler chronicles the genesis of the concept with whale watching as a recreational opportunity that stemmed out of conservation efforts to arrest the depletion of the Humpback Blue Whale in the 1960s (Ceballos-Lascurain 1996). Variously interpreted initially, Ceballos-Lascurain put it into perspective, thus,

> *"Traveling to relatively undisturbed or uncontaminated natural areas with the specific objective of studying, admiring, and enjj.ng the scenery and its wild plants, as well as any existing cultural manifestations (both past and present) found in these areas". (Commonwealth Department of Tourism 1994).*

Over the years, changing attitudes by travellers, tour operators, communities and conservationists have fine-tuned ecotourism substantially. Later The International Ecotourism Society defined ecotourism as -

> *"Responsible travel to natural areas which conserves the environment and improves the welfare of local people."*

Currently, a general consensus has been arrived at on the basic elements of ecotourism. Ecotourism

- contributes to biodiversity conservation;
- sustains well-being of local people;
- includes an interpretation/learning experience;
- involves responsible action on the part of tourists and the tourism industry;
- is delivered primarily to small groups by small-scale businesses;
- Requires lowest possible consumption of non-renewable resources; and
- Stresses upon local participation and ownership of business opportunities.

Ecotourism areas often include resident communities, specially of traditional people, and therefore, ecotourism plan must be holistic in approach and consider ways to conserve local cultural traditions and identities and how to bring benefits to them. Local populations are usually the first to be affected by any change in activities in their locations. Researches support community-based ecotourism (CBE) as a viable force for conservation by providing benefits to local people. Pragmatically managed, ecotourism can play a significant role in provisioning for livelihoods as an integral component of an ecosystem. However, meeting such conservation objectives requires careful and appropriate planning and design (Lindberg and Hawkins 1999). The WTO (1998) in formulating guiding principles towards sustainable ecotourism, strongly prescribe integration of local communities into tourism development by providing them jobs and income from tourism.

Community-Based Ecotourism

One of the essential ingredients of ecotourism is community participation since it claims to create opportunities for sustainable livelihoods. Local communities can play a significant role in the development of ecological and sustainable management of ecotourism module for community well-being. Thus, active involvement of the local community in tourism is imminent, specially since it is their domain for survival (Coltman 1989, WTO 1993, Cater and Lowman 1994).

CBE management refers to ecotourism programme, which take place under the control and active participation of the local people who inhabit or own a natural attraction (Lindberg *et al.* 1999). CBE works closely with local people, provides incentive for conservation, and more often, works outside of parks in locally owned land. It is a widely accepted truth that communities are an integral part of ecotourism system, because of which they are keen to sustain their ecosystem. CBE implies that a community is taking care of its natural resources in order to gain income through operating a tourism enterprise and using that income to better their lives. It involves conservation, business enterprise and community development. Community managed tourism consciously permits locals to decide the type of growth they would like to see and assists them to implement their plans (Lindberg and Hawkins 1999).

The Study Area

The Indian state of Madhya Pradesh (M.P.), located in the Malwa plateau, has immense unrealized potential for ecotourism development. The state is rich in natural and cultural resources of national and international repute, including the exotic Khajuraho temples, the Sanchi Stupa, wildlife sanctuaries and national parks, among numerous religious, cultural, historical sites. Bhopal, is the capital city. Situated in the immediate precincts of this urban sprawl is Van Vihar, the smallest National Park of India and one of most unique in the world. This national park was established in 1983, when approximately 445 hectares of degraded and denuded hillock was provided legal protection so that nature could sustain itself. Today, the Van Vihar National Park stands as a visible witness to nature's capacity to regenerate and revive. Kerwa is endowed with scenic and aesthetic values that attract approximately a hundred thousand visitors annually and is deemed to possess tremendous tourism potential. This green patch is in close proximity to Van Vihar National Park, which draws substantial crowds of visitors with its wide range of fauna and flora. Though Kerwa has a total area of barely 50 sq km, it attracts attention due to the following reasons:

- Location - scarcely thirteen kilometres from the main centres of the capital;
- Wee-wilderness - wilderness experience for nature lovers;
- Habitat - naturally endowed with serene forests and wildlife;
- Water resource - including a dam, perennial water stream besides smaller water bodies;
- Topography - interesting and aesthetic landscape, suitable for trekking.

The Van Vihar National Park and Kerwa forested area are situated close to each other, with the former designated as a zoo while the latter falls into the category of revenue forests, part of which is under the jurisdiction of the Forest Department. The present study was motivated by the growing interest and number of visitors to the Kerwa and Van Vihar areas. An interesting fact is that most of these visitors are domestic tourists, including villagers from adjoining rural areas. A tourism potential survey was conducted to gain insight into the tourism resource of the area and to assess the potential stakeholder's interest in combining efforts for sustainable CBE in the location.

Objectives

Based on the premise that small towns and rural areas have considerable potential for ecotourism (Gunn 1988), specific objectives for the study were determined, as follows.

- Identify key emerging issues in CBE, specially the community benefits
- Generate options to address the key issues

The study was intended for the following outcomes.

- Suggest approaches for short-term and long-term strategy for CBE
- Develop an operational module for CBE as a pilot project in Kerwa
- Formulate recommendations for state ecotourism policy guidelines

Method

The study was accomplished through participatory approach at all levels involving as many stakeholders as possible to ensure their active participation. Stakeholders included various levels of officials from the departments of Forest, Tourism, Energy and Rural Development; tourist operators; NGOs; professionals; scientists and academicians; community members; village level institutions including forest committees and *Gram Sabha* (village governance) members.

The methodology included intensive and extensive discussions, focus group discussions, village and house survey through schedules, field visits by combined team of major stakeholders followed by brain storming sessions and presentation of results of the analysis and open house discussion in a workshop.

The study was conducted between October 2000 to March 2002. Field surveys were carried out in all the seasons and responses were elicited from resident as well as the visiting population. The sample size were as follows;

- 40 households - households were selected through random selection, and
- 69 tourist groups (approx. 500 individuals) - Group size ranged between 6 and 10

It must be noted that though the questions were addressed to the heads of the groups/households, conscious efforts were made to obtain responses from all group members, to ensure a participatory approach. The findings, thus obtained, were discussed in a workshop, participated in by the major stakeholders.

Table 1. Facts of Villages Mendora and Chichli Profile

Description	*Mendora*	*Chichli*
Number of families	64	32
Population	650	187
Total agriculture land (Acres)	Nil	77
Livestock	80	191
Distance from ecotourism destination area	2 Kms The village is on way to the area	3 Kms This village is situated on the other side of the area
Village area	8 acre	5 acre
Communities	3-ST,SC, General	2-ST, SC

Source: Summary of Data Collected from Villagers

Villagers' Profile: General

For the purpose of villagers' perception two adjoining villages named Mendora and Chichli were selected, since these were closest to the ecotourism destination area.

Traditional Economic Activity of the Area: The traditional businesses of the area are agriculture (30%), labour (30%), and animal husbandry (23.3%). Remaining people are involved in fishing (15.3%).

Present Occupation: Most of the villagers of these two villages are occupied in animal husbandry (72%), followed by nursery plantation (52%), agriculture (21%), cow dung selling (17%), *tendu* leaves collection (13%), fishing (11%) and others (3%).

Local Handicraft: Villagers feel that the 'broom', which is presently being made for self-consumption, can have potential as handicraft. Besides this, some ladies are also involved in *'Rangoli'.*

Existing Committee Conservation Activities: 'Mendora Village Forest Committee' contributes to the conservation of forest. The committee is involved in plantation (70.2%), fencing (15.7%), checking illegal felling of trees (8.5%), protection from fire (8.5%), and land filling and sanitation works (1.4% each).

Awareness about (Eco)tourism: Most of the villagers are aware about purpose of the visit to the area by visitors, thus they are aware of tourism in their own perception. Nearly 76% of them are aware of ecotourism and related activities, though this was understood to be nature-based tourism.

Perception of Visitations: Through the perception of their interaction with the visitors, the villagers consider that people usually visit the place for picnic and recreation (60%), to enjoy nature (48%), wildlife (13%), fairs and festivals (10%), adventure sports (3%) and for religious purpose (2%).

Awareness of Visitor Origins: As per the villagers' perception about 43.4% people are local people who come from in and around Bhopal. Approximately 42.2% were outsiders, who came from different places. It was interesting to note that foreign tourists (14.4%) also visited the area, and stayed with them as their guests.

Visitor Profile

Tourist Influx and Origin: Data indicates that rainy season is the peak period when the number of visitors to Kerwa is the maximum. These visitors show a special preference for Sundays and general holidays. This is followed by winters while the minimum influx is in summers. Nearly 79% of the people visiting Kerwa are local visitors from different parts of Bhopal while 21% are outsiders.

Table 2. Seasonal Patterns of Visitors

Seasons	*Average Sunday (annual)*	*Per day other holidays (annual)*
Rains (July/ August)	200 (16,000)	1000 (8,000)
Summer (May/ June)	500 (4,000)	125 (1,000)
Rest of the Months	1200 (38,400)	600 (19,600)
Average	1217	596
Total	58,400	28,600
Annual total		87,000

Purpose of Visit: The majority of visitors visit the area for recreation. The number of nature-loving visitors is also quite good. However, from the discussion it emerged that the dividing line in the perception of the picnic (57.8%), outing (17.5%), and love for nature (15.7%) was quite thin.

Length of Stay: Majority (61 %) of the tourists are willing to stay up to three days if adequate facilities are provided, such as safe drinking water, better roads, trekking, rock climbing, lodging, park, restaurant, medical facilities, and fuel.

Willingness to Pay for Stay: Majority of the tourists agree to pay up to Rs 200/- per person per day for the necessary facilities.

Frequency of Visits Per Month: About 91% people visit Kerwa between 1-5 times a month. About 6% people visit 6 to 10 times a month. Only a small percentage (1.75%) have visited it for more than 15 times in a month.

Visits to Adjoining Locations: Besides Kerwa there are some other places also in and around Bhopal, which are of tourists attractions. These places are Kolar Dam, Bhojpur, Bhimbetka, Halali Dam, Lakeview, Kalia Sot Dam, Manbhavan ki Tekri (the highest peak of Bhopal), Delabadi (wildlife area), Lalghati, Chiklod, Raisen Fort and Indira Gandhi Dam. Percentage of the visitors who also visit these areas regularly are as follows.

Table 3. Neighbourhood Visitors' Attractions

S.No	*Name of Place*	*Percentage*
1.	Kolar Dam	40.7
2.	Bhojpur	33.3
3.	Bhimbetka	25.9
4.	Halali Dam	25.9
5.	Lakeview	14.8
6.	Kalia Sot Dam	14.8
7.	Manbhawan Ki Tekri	7.4
8.	Delabadi	7.4
9.	Lalghati	3.7
10.	Chiklod	3.7

Education Level of Tourists: Most of the people visiting Kerwa are educated. The result of the survey indicates that 37.04% group heads are postgraduates while the percentage of graduates is 33.33%, followed by 18.52% matriculates and 11.11% technical.

Occupation and Income Groups: The survey revealed that around 85% group heads belong to the service class, of which 15% were from technical field while the rest were non-technical. Of the remaining fifteen per cent, 11% are from business community and 4% are students. Most of the people visiting Kerwa belong to the middle class earning a monthly income ranging between Rs 3,000 and Rs 10,000, followed by 26% between Rs 10,000 and Rs 15,000. The percentage of visitors whose income lies between 1500120000 is 19%, while that of between 20001-30000 is 15%. A very small percentage of people visiting Kerwa is from the high income group.

Problems and Difficulties Encountered During Visit - Visitors complained about non-availability of safe drinking water (64.5%), poor public transport (22.8%), and non-availability of general stores (8.7%). Also, no trained guide is available for proper guidance about the area. There is no arrangement to check the visitors taking a dip or a swim in the dam for pleasure and fun, as a result there is always a fear of accident. Every year there are reports of few casualties. Besides, the visitors also face the problem of proper toilets and health care facilities.

Attitude Towards Locals - A large percentage of visitors (74%) expressed willingness to stay with the villagers. Visitors showed lot of interest in local culture including local folk songs and dances. Visitors were also found to be interested in local cuisine.

Resident-Tourism Interface

Economic Benefits from Tourism: It has been observed that only a small percentage of local people benefit from tourism. The main beneficiaries are petty shopkeepers e.g. grocery stores, tea stalls.

Willingness to Host Tourists: About 72% people are in favour of hosting tourists, while 25% are not in favour of tourism. Some 3% per cent people did not respond.

Willingness of Community to Control Tourism - Nearly 75% of the villagers have been found to be interested in community control of local tourism. They are of the opinion that village level forest committees could take over the activities of the tourism.

Negative Reactions: Villagers do not like tourists' behaviour to spread dirt and wastes anywhere on roads and other places, thus spoiling the area by dumping empty bottles and waste food in water, resulting in garbage and pungent smell. Boisterous behaviour, particularly among the youth, was observed to be on the increase because of tourism. Careless drinking and subsequent intoxication was specially noted. However, few women complained against the tourists entering their boundaries, and causing negative influence on children.

Locals' Interest in Ecotourism: About 75% of the villagers are willing to be associated with ecotourism and related activities. They are of the opinion that this may improve their economic status as they will get more opportunities to earn through tourism. Showing interest in various enterprises and earning activities, about 23% people want to establish small shops, 23% want to become tourist guides, 19% are in favour of establishing hotels or restaurants, followed by trading (10.7%), general employment (8%), self-employment (5%), folklore and handicraft (2.2%), cottage industry (1.1%) and adventure sports (1.1%). Significantly, as against the opinion of men, the women appeared apprehensive that their children might be affected by undesirable behaviour and lifestyle of some of the tourists.

Villagers were interested in ecotourism ventures if launched at the village level so as to enable their young people to get better job opportunities. Besides men, women may also earn through home-run and family-owned enterprises such as preparation of food *(papad* and pickles, local cuisine), tailoring and stitching.

Community Expectations and Concerns

The locals, however voiced concern with respect to infrastructural requirements to enable visitor experience. The villagers of Mendora and Chichli villages reported that in order to adopt ecotourism for enhancing their economy, they need certain basic facilities for tourism development. About 18% wanted infrastructural facility, followed by drinking water (12.5%), waste management and disposal facility (9.4%), electricity (9.4%), roads (9.4%), transportation (8.9%), communication (8.0%), hoardings (6.7%), men and women toilets (6.7%), police station (4.9%), children park (3.1 %), speed breakers (2.2%), medical facility (0.4%).

Capacity Building: Villagers appeared quite clear that to take up ecotourism as an economic venture they require capacity building specially for

- entrepreneurship development (self employment ventures);
- tourism activities (guides/ management); and
- small scale industries like making of *papad,* pickles, basket weaving and stitching.

Financial Support/Subsidies: The villagers also consider that some initial financial help is essential to invest in small-scale industries, shops, dairy, livestock, raw material, starting business, etc.

Infrastructure Development: Villagers feel that some infrastructure development is required for operation of the whole tourism activities especially sheds and other logistics.

Institutionalization and Regulations: The villagers firmly believe that in order to institutionalize the process, some regulatory measures are required. Institutional mechanism is considered necessary for providing logistics, support services, and indigenous fuel supply.

Suggestions by Village Community

- Hoardings at prime locations should be raised for drawing the attention of visitors towards tourists' code of conduct and better civic sense.
- Efforts should be made to educate tourists about proper disposal of waste in order to foster eco-friendly atmosphere and clean environment.
- A checkpost at the entry and near the dam be established to monitor anti-social, undesirable and accident possibilities.
- Entry fee may be imposed.
- Fishing in the dam should be allowed so that the local community may be benefited.
- Publicity campaign be launched to enlighten the tourists as well as the villagers about environmental laws, rules, and regulations framed by the government from time to time.
- Banners like "Make Your Trip Green and Clean", "Wandering one Gathers Honey" "Enjoy Nature — the Glory of God" be placed all over the area.

Conclusions

The area has adequate potential for small-scale ecotourism development and with appropriate and organized planning and implementation, conservation and economic development of the area is possible. The CBE module can be implemented as the local people are aware and sensitive about the tourism ventures. Thus, they can be realistically involved in the planning and implementation of ecotourism. Most villagers strongly feel that ecotourism should be encouraged. Perhaps, the public sector should take a lead. While there is appreciation for tourists, there is a strong resentment against the non-environmental tourist behaviours. However, there is a need for capacity building that will provide them opportunities for starting business at small-scale level. The youths of the village, in particular, are keen to be a part of ecotourism to act as tourist guides. Women are also interested to participate in ecotourism related activities.

There is need to streamline the functioning of the village forest committee with a view to involving villagers in the development and management of the area. An integrated holistic plan for community-based ecotourism should be prepared through the involvement of all stakeholders and implemented through the local people.

References

Cater, Erlet and Gwen Lowman (Eds) (1994). *Ecotourism: A Sustainable Option?* London. John Wiley & Sons.

Ceballos-Lascurain, Hector (1996). *Tourism, Ecotourism and Protected Areas.* Cambridge, UK. International Union for Conservation of Nature and Natural Resources (IUCN).

Coltman, Michael M. (1989). *Tourism Marketing.* New York. Van Nostrand Reinhold.

Commonwealth department of tourism (1994). *National Ecotourism Strategy.* Canberra. Australian Government Service.

Eagles, P.F.G. (1996). Issues in Tourism Management in Parks: The Experience in Australia. *Australian Leisure* 7(2): 29-37.

Figgis, P. (1993). Ecotourism: Special Interest of Major Direction? *Habitat,* 21 (1): 8 - 11.

Figgis P. (1994). *Ecotourism: Panacea, Plague or Possibility.* Eighth Richard Jones Memorial Lecture, October. University of Tasmania. Hobart. Central for Environmental Studies.

Gunn Clare A. (1988). *Tourism Planning.* New York. Taylor & Francis.

Harris R and N. Leiper (Eds) (1995). *Sustainable Tourism: An Australian Perspective.* Chatswood. Butterworth-Heinmann.

Lindberg Kreg and Deonal E. Hawkins (eds) (1999). *Ecotourism: A Guide for Planners and Managers.* Vol I, Dehradun. Natraj Publishers.

Lindberg, Kreg, Megan Epler Wood and David Engeldrum (eds) (1999). *Ecotourism: A Guide for Planners and Managers.* Vol II Dehradun. Natraj Publishers.

Kamra, Krishnan A. (2001). *Economics of Ecotourism.* New Delhi. Kanishka Publishers.

Nelson, J. G. (1994). The Spread of Ecotourism: Some Planning Implications. *Environmental Conservation* 21 (3): 248 - 255.

Sagasti F. and M. Colby (1993). Eco-Development and Perspective's on Global Change from Developing Countries. In N. Chourci (ed.) *Global Accord.,* London. MIT Press.

WTO (1984). *Economic Review of World Tourism.* Madrid. World Tourism Organisation.

W.T.O. (1991). *Yearbook of Tourism Statistics.* Madrid. World Tourism Organisation.

Wight, P. (1993). Sustainable Ecotourism: Balancing Economic, Environmental and Social Goals Within an Ethical Framework. *The Journal of Tourism Studies* 4(2): 2-10.

W.T.O. (1993). *Sustainable Tourism Development: Guide for Local Planners.* Madrid. World Tourism Organization.

W.T.O. (1998). *Guide for Local Authorities on Developing Sustainable Tourism.* Madrid. World Tourism Organization.

Ecotourism issues and options for New Zealand

Ross Cullen, Kay Booth and Ken Hughey

Natural sites are key attractions for tourists in New Zealand. Soaring international visitor arrivals and strong growth in domestic tourism is placing increasing pressure on many New Zealand nature-based tourism sites. A rapidly growing number of businesses advertise their activities to tourists as 'ecotourism' but the term suffers from ambiguity and lacks ownership. This paper reports ideas from a New Zealand workshop comprising representatives of the tourism industry, government policy agencies, conservation NGOs and researchers who recently debated the future of ecotourism in New Zealand. Specifically we report issues surrounding use of the term 'ecotourism', the policies and strategies required to maximize benefits from ecotourism including environmental benefits, and the actions required to verify the social, economic and environmental benefits which are claimed by ecotourism providers.

Tourism in New Zealand

Ecotourism is big business in New Zealand, a country which relies on its 'clean green' image to attract international tourists. Selling tourism experiences in natural areas to both domestic and international tourists is a plethora of tourism operators, mainly small in size, many of whom advertise their product as 'ecotourism'. Concerns about the loose use of the term 'ecotourism' and effects of nature-seeking tourists on the natural environment led to a national workshop on ecotourism in April 2002. The two-day meeting was organized by the Isaac Centre for Nature Conservation from Lincoln University, which holds an annual workshop to debate conservation-related issues. These are known as the Flock Hill Workshops, named after their location, an intermontane basin, 100km north-west of Christchurch in the South Island of New Zealand.

New Zealand tourism has grown rapidly from a small base for more than forty years, with international arrivals doubling in the last ten years (Statistics New Zealand 2002). Inbound international arrivals exceeded two million in 2001 (based on a resident population of 3.9 million). These visitors, when combined with domestic tourists, amount to 250,000 people travelling each day within New Zealand (Booth and Cullen 2001). Travel, and significant energy use, are essential components of much tourism within New Zealand (Becken and Simmons, 2002). Nature and natural features are major tourist attractions in New Zealand. The New Zealand land area is 270,000 km^2, and many tourists travel widely visiting the fourteen national parks and other natural attractions during their vacation.

New Zealand is almost unique in having zero charges for entry to national parks and reserves. Freedom of access and entry to rivers and coastline, national parks and most other protected areas is enshrined in New Zealand legislation. Continuance of that policy is supported by large segments of the population who regard freedom to interact with nature as a birthright.

Recent research has highlighted conflicting claims over the importance of ecotourism in New Zealand and Australia (Higham *et al.* 2001; Coventry 1997). Higham *et al's* research identifies 247 ecotourism operations nationwide, out of 400 nature-based tourism operators. Coventry (1997) argues that ecotourists comprise about 4 – 5 per cent of the New Zealand market.

The importance of nature in New Zealand tourism is very obvious to visitors and New Zealanders. New Zealand has a small population, no city larger than one million people, and limited numbers of cultural or historic features to attract tourists. Much tourism in New Zealand is focused on scenery and activities in natural settings. Ecotourism and voluntary codes of conduct are sometimes proposed as means to ensure that the negative effects of tourism on nature and natural settings are avoided or mitigated. The validity of those claims is not well established.

The Workshop

The workshop topic, 'Ecotourism: An ally of nature conservation?' was chosen in cognizance of 2002 being both the International Year of Ecotourism and the International

Year of Mountains. The belief of the organizers was that ecotourism is an increasingly fashionable label, but in danger of overuse to the point of becoming meaningless. There is a real possibility of overstating the virtues of ecotourism, and there is little New Zealand evidence documenting the reputed superiority of ecotourism over mass tourism. These issues were judged worthy of examination and debate. The stated intent of the workshop was:

- To consider whether ecotourism was indeed an ally of nature conservation.
- To ask what polices or strategies should be put in place to ensure that ecotourism is an ally of nature conservation.
- To determine how the positive benefits claimed for ecotourism could be verified.

Participation at the Flock Hill Ecotourism Workshop was by invitation with attendees from government (both local and national), conservation NGOs, industry, academia, and research organizations, to ensure that the full spectrum of experience, knowledge and expertise was focused on the topic. Attendees were asked to park their employment affiliations at the door and contribute frankly to the debate. To support that request, all Flock Hill Workshops are conducted under the Chatham House Rule that no comment expressed at a workshop can be attributed to any individual.

A background paper circulated prior to the workshop (Simmons 2002) outlined the three issues for discussion at the workshop (as noted above) and summarized New Zealand ecotourism research findings. The paper identifies that the term 'ecotourism' can only justifiably be used in stringently defined and verified cases and argues that ownership of the term ecotourism is needed to prevent continued overuse of the term. Certification schemes, it suggests, may be a means to accomplish ownership of the term while contributing to environmental protection and related goals.

The workshop commenced with a *powhiri*, a traditional Maori welcome to visitors, led by elders of Ngai Taihuriri, the local *runanga/tribal* council. The *powhiri*, the participation of Maori leaders and a Maori tourism operator, and a field trip to a local natural site with strong Maori cultural values all provided a platform for discussion of culturally-based tourism during the workshop. Cultural tourism was highlighted later in the workshop during a visit to a nearby *topuni* area (sacred place), Kura Tawhiti. This visit, with interpretation by a Ngai Taihuriri representative, dramatically exposed many participants to several of the issues and conflicts, including defacing of rock art and human waste entering waterways, which arise when tourists and recreationists visit culturally sensitive sites.

Short invited presentations were made at the outset of the workshop to raise issues and direct discussion. Presenters included a tourism operator, a tourism industry professional organization representative, a government tourism spokesperson and a government conservation spokesperson. Their arguments directed the structure and content of the workshop and can be summarized under the following points.

- The definition of, and criteria for, ecotourism is being established internationally. New Zealand should adopt the decisions projected to occur in Quebec at the World Ecotourism Summit in May 2002.
- In Australia and New Zealand Green Globe 21 and Qualmark schemes are being modified to provide the certification necessary to proscribe and verify ecotourism.
- A focus on ecotourism is too narrow - nature-based tourism is more useful. There is a spectrum from light to dark 'green'.
- There is a lot of marketing hype regarding ecotourism. The willingness of consumers to pay for certified ecotourism products, particularly when at a higher cost than similar products from conventional operators, was questioned.
- There has been a rapid growth in tourist numbers to natural sites and consequent pressures are being placed on some intensively-used sites. In some cases facilities may alleviate these pressures, in others they will not.
- New Zealand faces a choice in the types of tourism products it provides between having many certified ecotourism operations that cater for niche markets, or relying upon nature-based mainstream tourism operations.

Following the four initial presentations, workshop participants rapidly reached consensus that ecotourism is being defined internationally and that ecotourism operators must be able to demonstrate that they are ecologically, economically, and socially sustainable. Hence, fully functioning ecotourism operators should be able to readily justify claims that their operations are allies of nature conservation. Flock Hill participants asserted that ecotourism was a subset of nature-based tourism so the workshop would best examine the larger nature-based tourism sector. It was felt unlikely that ecotourism performance could be readily used as a benchmark to motivate better performance from mainstream tourism.

Ecotourism as an ally of nature conservation

The agreed vision from the group was: a tourism sector that supports a healthy environment. Participants recognized the need to develop strategies and policies to manage or moderate the negative effects of tourism and the workshop resolved to focus its energy on those issues. Two objectives were agreed upon:

1. To ensure that those in the tourism industry operate in accordance with ecotourism and nature-based tourism standards, attract customers and contribute to maintaining and improving the environment.
2. To ensure that the cumulative effects of the ecotourism and nature-based tourism industry on New Zealand's natural and cultural heritage are neutral or positive.

It was noted that in order to achieve these aims, the industry itself must remain financially self-sustaining.

Certification was discussed as a means for the industry to self-regulate its performance and ensure that it was 'triple bottom line sustainable' through meeting requirements such as Green Globe 21. It was argued that there are several advantages to certification including international recognition, marketing advantage, and cost-saving benefits to firms. Subsequent debate, however, noted that certification will only be adopted by some firms. The New Zealand tourism sector is renowned for its ease of entry and the numerous part-time members who are not certified currently and not members of a professional industry association. Thus certification alone cannot ensure that tourism is an ally of nature conservation.

In the absence of industry self-regulation, questions were raised whether the industry can be managed directly by limiting numbers or types of concessionaires or managing effects of the operations so that they do not exceed agreed environmental limits. Improved monitoring is needed for this purpose. Zoning could be used more frequently to determine which are the designated high-use, surface-hardened areas, and which areas are reserved for lower use and are less modified.

Tourism operators are legally required to obtain concessions from the Department of Conservation before taking clients onto any public conservation lands. It was argued that there are 10 – 20 hotspots in New Zealand. These sites should be identified, facilities developed to cope with high use pressures, and charges introduced to provide funding for management of these areas. Several participants later commented that it was surprising how often charges were suggested as a means to control use of sites and to collect revenue to fund their management. This is not so surprising when we recognize that the number of tourists has grown very rapidly, there is more pressure on resources, and tourist density is high and rapidly increasing in many smaller communities. The need to manage tourism better and the need for funding for these activities is much more apparent. Obstacles include anticipated strong public opposition to the suggestion of access charges and current legislation which does not allow charges for access to most state conservation areas.

A notable feature of the debate at Flock Hill was the apparent willingness of participants to propose a variety of controls on the tourism industry. The controls suggested included only allowing certified tourism firms to operate in the industry, increased monitoring of concessions, greater use of charges for access, and rationing the use of critical resources. From a business perspective the proposed controls are likely to bring increased compliance costs, reduce firms' profitability and hence be opposed by industry. One explanation for the apparent support for increased controls is the composition of the Flock Hill group. Participants included incumbents in the tourism industry, industry spokespersons, and government representatives. Particularly, members of the first two groups may have a vested interest in regulating entrance to the industry, and behaviour of firms within the industry. Newcomers are more likely to be excluded if there are limits on numbers of concessions available for key sites, resulting in higher profits for incumbents. Ease of entry to the tourism industry, overuse of some sites, and low financial yield are characteristics of the New Zealand tourism industry. Policies which restrict entry to the industry or to key sites, will benefit incumbents.

These two approaches — certification and regulation— it was argued, could be supported by education actions, better support by local government in establishing and enforcing resource management policies, and by the positioning of New Zealand in the marketplace. Education could play a significant role in improving industry performance if tour guides are required to be well trained and alert to issues of timing, location and content. The Department of Conservation and the New Zealand Tourism Industry Association could both play lead roles in supporting education initiatives.

A participant offered some reflections following the visit to Kura Tawhiti mentioned above. He argued that *taonga* (treasures) such as rock art must be better protected than they are at present, particularly if we allow tourists uncontrolled access to *topuni* sites. He pointed to the need for much better information supply particularly for tourism operators, about the importance and cultural significance of *topuni* sites. Tourism operators must be prepared to engage in dialogue with tribal groups to ensure that cultural values were not trampled by tourism, and sincere dialogue would result in mutually acceptable outcomes being achieved.

New Zealand could reinforce the attempts to self regulate, regulatory actions and education actions by smarter marketing policies. Marketing strategies should be sensitive to capacity levels and management requirements in regions and sites. Marketing awards should highlight genuine ecotourism operators allowing wholesalers and retailers to promote 'go with the local experts'.

Final Thoughts From Flock Hill

Tourism is a young industry in New Zealand. Tourism has grown very rapidly, and the pressures it exerts particularly on society and parts of the environment, have become noticeable as visitor numbers have doubled and redoubled. The vision espoused by the workshop group - a tourism sector that supports a healthy environment - is worth striving for. A combination of industry self regulation (certification), concessions limitations, more astute involvement by local government, effects monitoring and regulation, buttressed by education and marketing strategies will take New Zealand some way to achieving this vision for nature-based tourism.

References

Becken, Susanne and David Simmons (2002). Understanding Energy Consumption Patterns ot Tourist Attractions and Activities in New Zealand. *Tourism Management* 23: 343-354.

Booth, Kay and Ross Cullen (2001). Managing Recreation and Tourism in New Zealand Mountains. *Mountain Research and Development* 21(4): 331-334.

Coventry, Nigel (1997). Ecotourism. *New Zealand Chartered Accountants Journal* May: 33-35.

Higham, James, Anna Carr and STEPHANIE GALE (2001). *Ecotourism in New Zealand: Profiling Visitors to New Zealand Ecotourism Operations.* Research Paper No.10. Department of Tourism. University of Otago. Dunedin.

Simmons, David (2002). *Eco tourism: An Ally of Nature Conservation? Defining the Rules and Measuring the Outcomes.* Flock Hill 2002 Proceedings. Lincoln University.

Statistics New Zealand (2002). *Key Statistics*, March.

Ecotourism in the Australian and New Zealand sub-Antarctic islands

C. Michael Hall

The development of special interest tourism, with particular attention being placed on the adventure, educational, and environmental market segments and advances in polar technology has been a growing awareness of the potential for a substantial growth in the levels of Antarctic tourism. In addition, the possibility exists for commercial tourism operations to subsidise government scientific operations and justify claims to Antarctic and sub-Antarctic territory (Reich, 1979; Hall 1990, 1992a; Beck 1990a,b). However, tourists have disrupted scientific programmes through either their presence or the need to offer emergency services, and have created substantial legal problems concerning jurisdiction and liability. Indeed, tourists now exceed the number of government personnel (Stephenson, 1990), although the period of time they spend in the Antarctic region is substantially shorter. Furthermore, tourism has the potential to dramatically impact the fragile sub-Antarctic island environments, a situation which has already occurred in several islands and on the Antarctic continent (Lyster, 1984; Hall, 1992a). This paper examines the implications of ecotourism for the sub-Antarctic islands of Australia and New Zealand, a region which is increasingly seen as offering a model for Antarctic and sub-Antarctic tourism management (Radio New Zealand, 'Morning Report', 1992). The paper is divided into four sections: First, an outline of the nature of sub-Antarctic tourism; second, the impacts of tourism; third, the management of tourism in the Australian and New Zealand sub-Antarctic islands, and finally, a discussion of the implications of sub-Antarctic ecotourism for notions of sustainable tourism.

Sub-Antarctic Tourism

In the Antarctic and sub-Antarctic context, tourism is defined as 'all existing human activities other than those directly involved in scientific research and the normal operations of government bases' (Hall, 1992a). This definition covers the activities of commercial tourism operations, non-government expeditions and the recreational activities of government personnel. Tourist activity is primarily concentrated to cruiseship and flight operations in the region of the Antarctic Peninsula, the Ross Sea and associated sub-Antarctic islands. However, in recent years increasing interest has been given to the commercial tourism potential of the Australian and New Zealand sub-Antarctic islands and associated Antarctic territories (Hall, 1992a).

A survey of the interests and activities offered by operators through the *Specialty Travel Index* between 1987 and 1990 reveals that the most popular touristic activities undertaken in Antarctica are cruising, flightseeing, mountaineering and nature oriented travel (Hall, 1992a). Since the first tourist flight to Antarctica in 1956 the level of visitation to the region has remained relatively static until recent years (Reich, 1980; Codling, 1982; Wace, 1990; Enzenbacher, 1992; Hall, 1992a). Cruises are the most available and therefore most popular form of tourism with between 2,500 and 3,000 people visiting the Antarctic region each year over the past decade (Reich, 1980). However, one prediction for the 1988-89 Austral summer gave an estimate of 7,000 tourists (Stephenson, 1990). According to the blouse of Representatives Standing Committee on Environment, Recreation and the Arts (HRSCERA, 1989:2), 'Patrons of organised commercial Antarctic operations tend to be affluent older people, with potential medical/fitness problems; the average age of the 80 passengers on one cruise was 60 years'. Nearly all cruises begin from Punta Arenas (Chile), Puerto Williams (Chile), or Ushuaia (Argentina). However, as well as visiting the Antarctic Peninsula, the majority of cruises will also visit sub-Antarctic island groups such as South Georgia, South Sandwich islands, South Orkney Islands, South Shetland Islands, and the Falklands.

Tourism landings are not permitted on the South African Prince Edward and Marion Islands, while commercial tourist interest in the French He Amsterdam, lie Saint-Paul, Iles Crozet, and Iles Kerguelen is low, although the rate of visitation to the Australian, British, and New Zealand's sub-Antarctic islands has increased considerably in recent years. In the case of both the continental Antarctic and the sub-Antarctic islands, the most prominent issue surrounding tourism is the potential environmental impacts of tourism. Although the impact of tourism may be insignificant compared to damage created by national scientific bases, the harvesting of marine life, or oil and mineral exploration and extraction, it still remains as one of the main barriers to the development of the sub- Antarctic islands as tourist destinations (Kriwoken &Keage, 1989b).

The Environmental Impacts of Tourism

> ... some ecotour operators admit that no matter how carefully expeditions are run, the very presence of humans can affect a fragile environment. A footprint in the Antarctic moss could last for decades, a plastic pen dropped on to an isolated beach could remain there for centuries and even a minor oil spill could upset an ecosystem and kill off a rare species of plant or animal. The more ecotourists there are, the more likely it is that the environment being visited will be damaged. Ecoholidays, like mineral and scientific expeditions, will slowly and unwittingly consume one of the planet's most scarce, valuable and irreplaceable treasures: the time capsules of life that remain unpolluted by humanity (Masson, 1990:56).

Undoubtedly the most serious concern surrounding tourism in the sub-Antarctic islands is the potential adverse impacts of tourism on the physical environment. The sub-Antarctic islands contain some of the world's least human impacted biotas. Their relative isolation from population centres and their extreme climatic and physical conditions has been their greatest conservation asset. Nevertheless, it is the same harsh conditions and corresponding natural beauty which is now attracting visitors in ever-increasing numbers. However, their fragility means that even minute changes brought about through human impacts, such as tourist activity, may have major long-term impacts on ecosystem stability (Chittieborough, 1984; Nicholson, 1986; Hall, 1992a).

Island ecosystems are substantially different from those of continental areas. They are characterised by limitations of space, restricted habitats, lower species numbers compared to continental areas of similar ecological diversity, and a high degree of species endemism. 'The islands' biota is a culmination of a long history of geographical isolation, species dispersal, climatic factors, and community interaction— until very recently without human interference. The evolution of the island biota is of great international taxonomic and ecological interest, and the islands are of immense value for scientific study' (Higham, 1991:10). The specialisation of island biota means that they are highly vulnerable to external disturbance and environmental change, particularly human-induced impacts, and the extinction of species is particularly common on islands when new competitors or physical conditions are introduced. However, the insular characteristics of islands also underline their value as protected areas:

> Paramount among these values is the uniqueness of flora and fauna due to the presence of endemic, relict and/or specialised species. Their isolations means that islands are ideally suited as refugia for threatened plants and animals and as reservoirs for the preservation of genetic resources. Islands also offer much scope for the study, understanding and appreciation of intact and holistic natural ecosystems (Clark & Dingwall, 1985: 3).

Table 1. Environmental Impacts of Tourism in Antarctica

Type of Activity	Nature of Impacts	Infrastructure Characteristics
Overflights	fallout from engines; disturbance of wildlife due to noise	no requirement for permanent land-based facilities
Ship-based	transient environmental effects, although pressure may be placed on regularly visited land attractions; oil spills; disturbance to wildlife; potential introduction of bird and plant diseases; introduction of exotic flora.	no requirement for permanent land-based facilities
Onshore-facilities	increase demand for ice-free land and fresh water supply; disposal of sewage and rubbish; degradation of specific sites with high visitation levels; disturbance to wildlife; potential introduction of bird and plant diseases; introduction of exotic flora. Possible damage to cultural heritage sites.	Support infrastructure including the provision of an all-weather air strip capable of handling large commercial aircraft; accommodation facilities; potential combination of tourist facilities with scientific bases.

Source: Hall, 1992a

The characteristics of environmental impacts are closely related to the nature of tourist activity (Table 1). Overflights provide minimal disturbance of the environment and do not require permanent land-based facilities, although low overflights of wildlife colonies may panic the birds or marine mammals. For example, it has been alleged that a stampede resulting in the death of about 6,000 King Penguins on Macquarie Island in June, 1990, was caused by an overflight (*The Australian,* 25 June 1990). However, overflights are generally restricted to the Antarctic mainland.

Considerable debate exists over the impacts of ship-based tourism on the Antarctic and sub-Antarctic environment. The length of time that ship-based visitors spend on shore and their activities requires further study, particularly as ship-based tourism is extremely hard to regulate. Furthermore, cruise travel in the Austral summer coincides with the peak breeding periods for many species and may disturb breeding sites. Ships can pollute over a wide area through indiscriminate disposal of wastes and sewage, while oil spills would have a major impact on the fragile shoreline ecosystems (Codling, 1982; Hall, 1992a). An additional threat is the introduction of exotic species onto the islands through either seed dispersal or the transfer of mammals such as rats and mice.

Permanent onshore infrastructure has had the greatest impact on the Antarctic and sub-Antarctic environments. The establishment of tourism facilities poses major problems of waste disposal, food and water supply, and the provision of accommodation. The development of land-based tourism operations necessitates the construction of communication and transport networks. King George Island in the South Shetland Group illustrates some of the dangers of concentrating development in the Antarctic. There are seventeen major facilities on the island including a Chilean airbase and the stations of eight countries, while the impacts of station construction has led to the revocation of Specially Protected Area status on two occasions (Headland & Keage, 1985). Chilean and Argentinian bases on the sub-Antarctic islands have already been developed with some facilities for tourism. However, no such facilities exist on the Australian, French, New Zealand, Norwegian, and South African Islands, particularly as scientists have been keen to minimise the potential disturbance by tourists of scientific research.

Although the impacts of tourism in the sub-Antarctic are predominantly associated with the physical environmental, substantial concern also exists over the conservation of cultural heritage. A number of early European sites associated with farming, sealing and whaling exist on the sub-Antarctic islands and several early exploration bases are of substantial historic significance (Stephenson, 1990; Department of Parks Wildlife and Heritage, 1991; Hall, 1992b). For example, Smith and Simpson (1987) called for the conservation and protection of 19th century sealers' sites in the South Shetland Islands.

The conservation significance of the sub-Antarctic islands has long been recognised (Carrick, 1960, 1964; Holdgate & Wace, 1961; Holdgate, 1970; Roberts, 1977; Clark & Dingwall, 1985). In a comprehensive review of the conservation requirements of Antarctica and the Southern Ocean, Holdgate (1970) identified three principal management objectives:

1. The general protection of scenic beauty and the biota of the Antarctic region south of 60°S latitude.
2. The protection of remaining undisturbed ecosystems of oceanic islands north of 60°S, and as far as possible the restoration or stabilisation of those ecosystems that have been disrupted by actions of man.
3. The wise management of the biological resources of the Southern Ocean, to enable a sustainable harvest to be taken.

The first two objectives are particularly important for the management of tourism in the sub-Antarctic islands. The isolation of the island from population centres can no longer be regarded as adequate protection for the islands. Tourism has become a reality in the management of the island ecosystems. As Clark and Dingwall (1985: 179) recognised in their report for the International Union for the Conservation of Nature and Natural Resources (IUCN), 'It is unrealistic to lock-up the southern islands exclusively for meteorological or scientific purposes. Tourism has a valid place in the Southern Ocean, as long as it is regulated and carefully supervised'. The problem therefore becomes one of developing appropriate tourism management strategies, a concern which has focussed considerable attention on Australian and New Zealand attempts to manage visitation to their sub-Antarctic islands.

The Management of Tourism in the Australian and New Zealand Sub-Antarctic Islands

Australian and New Zealand interest in sub-Antarctic tourism is of relatively recent origin. In the absence of organised and regular tourism to their Antarctic and sub-Antarctic territories, both countries took a generally neutral position on tourism until the late 1980s (Bergin, 1985; HRSCERA, 1989). However, their leading role in Antarctic conservation, Treaty negotiations, and the maintenance of an Antarctic presence has led to the first steps being taken towards the development of a comprehensive tourism management strategy (Hearder, 1989; Hall, 1990).

Australia possesses three sub-Antarctic island groups (Heard Island, MacDonald Islands, and Macquarie Island) and New Zealand five (Antipodes Islands, Auckland Islands, Bounty Islands, Campbell Islands, and Snares Islands). The features of these islands are outlined in Table 2. In a biogeographical sense the Australian islands lie within the 'Subantarctic zone' identified by Clark and Dingwall (1985) while the New Zealand islands may be classified as 'Cool temperate' (Appendix 1). As with all the sub-Antarctic islands, the Australian and New Zealand islands show a high degree of endemism, host a number of rare and endangered species, and have significant breeding colonies of marine birds and mammals. Although the Maori visited the Snares Islands and possibly other New Zealand sub- Antarctic islands for food-gathering purposes, significant human exploitation of the sub-Antarctic islands only began with the arrival of European sealers and whalers at the end of the eighteenth century. The rapid decimation of seal and whale stocks, from which many species are only now recovering, meant that the direct human imprint on the islands was only shortlived, although the exotic plant and animal species they introduced, such as rats, mice, and rabbits, have persisted on many of the islands. Nevertheless, compared to many continental regions, the islands are virtually pristine. For example, 'Adams and Disappointment Islands in the Auckland Island group, and the whole Snares group are among the last substantial land masses of the world harbouring vegetation unmodified by plants and foreign animals' (Higham, 1991:11).

Table 2. Australian and New Zealand Sub-Antarctic Islands

Island	Area (ha.)	Maximum Altitude (m)	Physical Features	Flora and Vegetation	Fauna	Cultural Features	Management
Australia							
Heard Island	38,000	2,745	Dominated by the Big Ben massif, with the volcanically active Mawson Peak. Over 90% of area is glaciated, with ice up to 150m deep.	Eight species of vascular plants. There are no trees. No alien plants are present.	Large populations of southern elephant seal. Leopard seal and Antarctic fur seal numbers are increasing. Twenty species of bird breed on the Island (two endemic). No alien animals are present.	Discovered in 1833. Sealing gangs occupied the Island continuously from 1855-1875 and sporadically from 1876- 1929. Several whaling and sealing sites can still be seen. An ANARE Research Station was established in 1947.	Australian External Territory under Common-wealth control. A management plan is under preparation although guidelines for visits have been established. The Island is subject to the Convention for the Conservation of Antarctic Marine Living Resources. The Australian Government has nominated the Heard and MacDonald Islands to the World Heritage Convention.
MacDonald Islands	260	230	Islands composed of volcanic material resulting from eruptions of volcanic vents near sea level.	Five species of vascular plants.	Similar to Heard Island. Seal from the Islands probably recolonised Heard Island after over-exploitation of stocks in the 1800s.	Discovered in 1854. Exploitation of seal and penguin stocks throughout the 1800s. No alien plants and animals.	As fer Heard Island.
Macquarie Island (includes Judge and Clerk Islands, and Bishop and Clerk Islands)	12,785	433	Rocky shoreline with sleep cliffs, Island is volcanic in origin. Glacial activity in past but there is now no permanent ice.	No trees, but the Island is heavily vegetated: 40 vascular, 50 moss, 30 liverwort, and 55 lichen species. Three endemic vascular species. High degree of endemism among lichens.	21 breeding species of birds. Only known breeding ground of the Royal Penguin. A number of seal species also breed on the Island. There are several hundred species of invertebrates.	Discovered in 1810. Inhabited periodically by sealers and whalers throughout the 1800s. Several scientific expeditions visited the Island during the 19th century. A meteorological and scientific research station was established in 1948. Several significant sealing sites remain. Introduced animals include grey duck, mallards, wekas, cats, rats, mice and rabbits.	The entire island is declared a nature reserve under Tasmanian State legislation: 1933 — declared a Nature Sanctuary, 1972–declared a State Reserve, 1977 — declared a Biosphere Reserve, 1978– classified as a Nature Reserve. Entry to the Reserve is by permit. A detailed management plan was established in 1991. Limited tourist visitation is permissible under strict supervision.
New Zealand							
Antipodes Islands	2,100	402	Islands are of volcanic origin.	Recorded flora include 46 species of flowering plants, 18 pteridophyte, 23 bryophyte, and 15 moss species. 4 plant species are endemic.	25 species of birds breed on the islands. There are four endemicspecies of birds. There is a high degree of endemism among land invertebrates. Southern elephant seal is the only indigenous breeding mammal.	The islands were discovered in 1800. Sealing occurred until the 1820s. A castaway depot was maintained from 1886 to 1927. The island was leased for farming in 1895 but was never stocked. Mice are the only introduced fauna.	The entire island is a National Nature Reserve: 1961 — declared a Reserve for Preservation of Fauna and Flora, 1975 — foreshores were added to the Reserve, 1977 — designated a Nature Reserve, 1986 — National Nature Reserve. National conservation legislation also pertains to the Reserve. A draft management plan has been prepared for the Reserve. Visitation is strictly controlled.

(Contd.)

Auckland Islands	62,564	667	The group consists of a main island and several islets and stacks. The islands are the eroded remains of two basaltic volcanoes. The islands have been subject to several periods of glaciation.	Vascular flora is extremely rich and consists of 228 species inducting 44 species of ferns. The main Auckland Island is dominated by southern Rata forest.	The islands have a rich birdlife, with at least 46 species of breeding birds, including 8 endemic species. The islands are among the major breeding grounds in the world of the shy and wandering albatross. The islands are the main breeding grounds for Hooker's sea lion, several other seal species also breed on the foreshore. Several hundred species of terrestrial invertebrates occur, with a high degree of endemism.	The islands were discovered in 1806. Sealing and whaling occurred throughout the 1800s. The main island was inhabited by a small group of Maori from 1842-1856. A settlement (Hardwicke) was established in 1849 but was abandoned In 1852. Later attempts at sheep farming failed. Introduced fauna and flora occur on several of the islands, although an active eradication programme is underway.	The Auckland Islands are a National Nature Reserve: 1910 – Adams Island declared a Reserve for the Preservation of Fauna and Flora, 1934–the rest of the islands were included, 1975 – the Reserve was extended to cover the foreshore, 1977- classified as a Nature Reserve, 1986 –declared a National Nature Reserve. A management plan has been established for the Reserve which permits limited visitation, although tour parties must be accompanied by a ranger.
Bounty Islands	135	88	Comprises over 20 small granite islands, islets, and rocks.	No terrestrial vegetation has been described, although lichen and green algae reportedly occur.	The islands support the largest known breeding ground of New Zealand fur seal. Seven species of birds breed, including two endemic species.	The group was discovered in 1788. Sealing operations commenced shortly after until the decimation of seals by 1830. A castaway depot was established in the 1880s.	The islands are a National Nature Reserve: 1961 - Reserve for the Preservation of Flora and Fauna, 1975– foreshores added to Reserve, 1977 – classified as a Nature Reserve, 1986 – declared a National Nature Reserve. A draft management plan has been prepared with visits bring strictly controlled.
Campbell Islands	11,331	567	Campbell Island is a remnant of a dissected volcanic dome. The island was glaciated in the Pleistocene.	The vascular flora consists of 218 species, sub-species, and hybrids, with 119 species of moss.	29 species of birds breed on the islands, with one endemic. This islands support large beeding colonies of albatross and yellow-eyed penguin. Southern elephant seal, New Zealand fur seal, and Hooker's sea lion breed on the islands. Southern Right Whale congregate near the islands for breeding.	The islands were discovered in 1810 with sealing continuing until the 1830s when stocks were exhausted. Whaling ceased in 1916. Sheep farming from 1895-1931. A meteorological station currently operates on the island.	The islands are a Nationa1 Nature Reserve: 1954 — Reserve for the Preservation of Fauna and Flora, 1975— foreshores added to the Reserve, 1977 — designated a Nature Reserve 1986 — National Nature Reserve. Visitation is permitted but is strictly regulated. A management plan has been developed for the Islands.
Snares Islands	328	152	Snares Islands comprise two groups of islands and are composed of jointed granite.	Flora of the islands comprises 20 vascular, 27 moss and bryophyte, 6 fungi and 45 lichen species. The vegetation is dominated by Olearia forests.	The islands have 23 species of breeding birds with several endemic species. 2.75 million burrow-holding pairs of Sooty Shearwater were estimated to live on the islands in 1982. Several marine mammals breed on the islands.	There is evidence of a pre-European Maori presence on the islands. European discovery occurred in 1791. Sealing occurred throughout the 1800s. A castaway depot was established in 1867 and maintained until 1929. A research station was established in 1961.	The Islands are a National Nature Reserve: 1961 — Reserve for the Preservation of Fauna and Flora, 1975—extended to include foreshore, 1977 —designated a Nature Reserve, 1983 — designated a National Nature Reserve. A management plan for the Reserve has been prepared. No tourist landings on any of the islands is allowed, although cruising may be permitted.

Sources: Keag e, 1982; Department of Lands and Survey, 1983, 1984, 1987; Clark and Dingwall, 1985; Department of Parks, Wildlife and Heritage, 1991; Higham, 1991.

The wilderness character of the Australian and New Zealand sub-Antarctic islands has led to substantial commercial interest in developing the tourism potential of the islands both on their own and in conjunction with visits to the Australian and New Zealand Antarctic Territories. Some 564 people visited the Island in the 1990-1991 summer. There were no tours in 1991-92 and this situation may continue for the next three Austral summers. However, the limit will probably continue to stay at four ships and approximately 600 people per Austral summer.

New Zealand Sub-Antarctic Islands

All the New Zealand sub-Antarctic islands have the status of a National Reserve. This status requires an Act of Parliament to alter any conditions pertaining to the reserve. Management plans have been prepared for the Campbell (1983), Auckland (1987) and Snares Islands (1984) by the New Zealand Department of Conservation and their predecessor, the Department of Lands and Survey. The Snares Islands are not open for tourism, while a quota of 500 tourists per season (presently under review) was permitted for the Campbell and Auckland Islands in 1990/91 (Peat, 1991). A fee of NZ$100 per person is charged which is directed towards management programmes «such as the construction of over a kilometre of board walkon Campbell Island to allow tourists to visit an Albatross colony with minimal impacts on the environment.

The tourism policy for the Auckland Islands is 'To permit visits to selected areas of the reserve by tourists but under such controls as deemed necessary to ensure protection of its natural features, ecosystems and cultural values' (Department of Lands and Survey, 1987: 14). A similar policy exists for the Campbell Islands (Department of Landsand Survey, 1983:51-52). It is the intention of the Department of Conservation that both island groups will be managed according to '"wilderness" type visiting codes... visits are limited in number, the landings are supervised by a representative of the department, are under strict supervision and are of short duration' (Department of Lands and Survey, 1987: 14).

> Cruising expeditions must have a genuine educational or inspirational purpose relating to better appreciation of nature. Overseas tourist ships are required to have a current deratting exemption certificate as part of the permit to land on the islands.
>
> Often the most spectacular sight-seeing is obtained from the sea, and this activity is not restricted, (p.14)

The New Zealand sub-Antarctic island management plans have greatly influenced the management of Macquarie Island by the Tasmanian Department of Parks, Wildlife and Heritage (1991) which has adopted similar costing and management strategies, and guidelines for tourism operations (Stephenson, 1990).

Australian Sub-Antarctic Islands

Macquarie Island and the Heard and MacDonald Islands come under a range of Commonwealth (federal) and state legislation (Bergin, 1985). As an Australian external territory the Heard and MacDonald Islands come under Commonwealth control. A wide range of Commonwealth domestic legislation can be applied to the islands including the National *Parks and Wildlife Conservation Act 1975, Wildlife Protection (Regulation of Exports and Imports) Act 1982, Whale Protection Act 1980, Sea Installations Act 1987, Environment Protection (Impact of Proposals) Act 1974, Environment Protection (Sea Dumping) Act 1981,* and the *Protection of the Sea (Prevention of Pollution from Ships) Act 1983.* Under the *Australian Heritage Commission Act 1975* Federal ministers and authorities are required to refer to the Commission for comment any action which will affect a place in the Register of the National Estate to a significant extent. Both the Heard and MacDonald Islands and Macquarie Island are listed in the register. Although not subject to cruise ship visits, there have been a number of private expeditions to the Heard and MacDonald islands. Furthermore, the proposed nomination of the Islands to the World Heritage List may well encourage visitor interest. Therefore, it would seem imperative that the Common-wealth develop a management plan for the islands.

Macquarie Island is a Nature Reserve that comes under the authority of the Tasmanian *National Parks and Wildlife Act 1970,* under which a management plan has been prepared for the Reserve by the Tasmanian Department of Parks, Wildlife and Heritage (Stephenson, 1990; Department of Parks, Wildlife and Heritage, 1991). One of the objects of the Management Plan is 'to permit tourist visits under strictly controlled conditions which allow visitors to experience the natural values of the island without compromising them' (1991: 17). However, while the Plan (1991: 15) recognises that 'in the long term it is only with public understanding and support that the world's wildlife, habitats and natural ecosystems can be protected', tourism should only be encouraged insofar as it does not conflict with the Objects of Management, first of which is 'to protect and manage the reserve as a natural habitat for its indigenous flora and fauna and in order to achieve ecosystem conservation' (p.17). In order to achieve these goals the Prescription for Management states: 'Tourist visits will be ship-based but limited facilities such as walkways, viewing platforms and interpretation material may be provided in selected areas to protect the wildlife, environment, historical and/or scientific values of the reserve' (1989: 23). In addition to setting guidelines for the protection of scientific programmes and the safety of visitors and personnel, the *Guidelines for Tourism Operations at Macquarie Island Nature Reserve* set nine directives for the protection of the environment:

* All tourist operations will be ship-based with no overnight stay on the island except in an emergency. Shore visits will only be permitted between the hours of 0700 and 1900 local station time.
* The landing and pickup of personnel will only be at beaches designated by the Department.
* The areas which may be accessed on foot will be designated by the Department and all shore parties are to be in two-way radio communication with the ship and must not be more than one hour walking time from the beach where they are to be picked up.
* Shore parties to be organised in groups of no more than ten people including one leader/guide with each party.
* Strict quarantine procedures will be enforced to prevent exotic species being taken ashore in equipment or clothing.
* Any food and drink items to be consumed during visits ashore are to be unopened, pre-packed, processed food or drinks, previously approved by the Department.
* No food items are to be given to wildlife.
* All rubbish and unused food items are to returned to the ship. No shipborne rubbish, including food items, are to be disposed of in Tasmanian territorial waters.
* No collecting of flora, fauna, historical sites or artefacts, geological specimens or objects is permitted (1989: 51).

The success of the *Guidelines* in meeting both the conservation objectives and providing a satisfying tourist experience is still to be examined. Nevertheless, it should also noted that the Aus$100 charge per visitor under the Management Plan may place substantial economic burdens on smaller tourist operations and thereby further restrict tourist access. From a bio-physical perspective, the *Guidelines* may well be extremely appropriate for the management of visitation to sub-Antarctic islands and may also meet the requirements of the proposed World Heritage listing for Macquarie Island (Hall, 1992b). However, as Codling (1982: 7) commented, 'all forms of control, whether zoning or other management techniques, raise questions as to the timing of their introduction, effective enforcement and monitoring of their effect'. Nevertheless, the assessment of tourist activities would require a far more thorough consideration of Antarctic ecology and the relation to human impacts than has hitherto been the case.

Sustainable Tourism and Sub-Antarctic Ecotourism

From an ecological perspective, sustainable tourism means conserving the productive basis of the physical environment by preserving the integrity of the biota and ecological processes and producing tourism commodities without degrading other values. Having no form of tourism in the sub-Antarctic islands may well be the most advisable management strategy in terms of the ecological integrity of the islands. However, it is also unrealistic. In order to ensure that wilderness areas are preserved we must, somewhat paradoxically, allow people to visit these wild places so that policy makers can be persuaded to maintain their reserve status. Vicarious appreciation through books and documentaries is important, but it is not sufficient to create a groundswell of public opinion for preservation (Hall, 1992b). As Clark and Dingwall observed, 'With adequate precautions and international co-operation in regulating tourist operations, tourism should be compatible with scientific and conservation objectives in protected areas on islands of the Southern Ocean' (1985: 179).

Presently tourism is the only form of economic exploitation of the sub-Antarctic islands, although oil and mineral exploration, fishing and whaling all loom on the development horizon. Given that ecology alone is not sufficient to give wilderness value, tourism provides a useful economic argument to ensure that the sub-Antarctic islands are preserved in their relatively pristine state for future generations. Antarctic and sub-Antarctic tourism is important for several cities in Australia in New Zealand. For example, although Hobart is not an Antarctic tourism destination *per se,* the city is the staging port for travel to, and tourism in, the Australian Antarctic Territory and Macquarie Island. Hobart has strong links with Antarctica because of the number of international and domestic Antarctic-related agencies which are based there. In addition, the Tasmanian Government has indicated strong support for visitation to Macquarie Island (which lies under State control) and the establishment of an International Antarctic Centre in Hobart (Hall, *et al.*, 1991). Interestingly, the City Council of Christchurch, New Zealand's main staging post to the Antarctic, has publicly opposed Antarctic tourism, although it has been willing to promote itself as a centre to experience Antarctica at a distance Josefa, 1990: 16).

The management regimes which have been put in place in Australia and New Zealand are designed to ensure the sustainable use of the sub-Antarctic islands as a tourist resource and are seen as a potential model for the management of tourism on other islands and on the Antarctic. At their eighth meeting, Consultative Parties to the Antarctic Treaty acknowledged that Antarctic tourism is 'a natural development' (in Nicholson, 1986: 6). The establishment of a tourism management regime for islands over which national sovereignty is clear, is substantially easier than areas over which sovereignty is disputed. For example, in areas such as the Antarctic Peninsula where the majority of tourism activities are focussed. Furthermore, sovereignty allows for far greater direct regulation and control over the activities of both tourists and tour operators. However, the appropriateness of the Australian and New Zealand management models is still to be tested, particularly as some tourist operators have had little opportunity to contribute to the development of management strategies.

Davis (1984: 351) believed that tourist operators needed to have an input into the Antarctic policy-making process in order to establish effective management regimes and thereby ensure that development of the tourism resource is sustainable. Tourist operators must be encouraged to feel a degree of 'ownership' over resources in order to assist their resolve to manage and protect them. As Codling (1982:9) observed:

> It is in the interest of tour operators, who intend to return to the continent, to cooperate with the Treaty Nations, and they should be closely involved in any action taken to resolve pressures or conflict. Their own commercial interests are best served if their clients are satisfied, and there is value in seeking to understand and respond to visitor's requirements.

In Australia and New Zealand, the management of tourism in the sub-Antarctic islands has been based on concern over the potential impacts of visitation. The management regimes are strongly weighted in favour of ecological considerations. However, management agencies face a paradox, in order to maintain support for their management strategies they need to be able to give people access, albeit controlled, to the resource. Yet that access will impact, however minimally, the ecological resource. Management agencies therefore have to find a level at which ecological change is acceptable. In order to be able to do this they need to understand the impact of visitation on ecological processes and the effects of trampling and they need to be able to regulate the activities of tourism operations in a manner which does not threaten commercial viability. Unfortunately, at present it is too early to say whether these two major activities will be undertaken. While the policy and the management regime are in place, the effectiveness of the implementation process has still to be seen.

Acknowledgement

The author would like to gratefully acknowledge the assistance of Treve MacCarthy, Simon McArthur, and Mariska Wouters in the preparation of this paper.

References

Beck, P.J. 1990a Antarctica enters the 1990s: an overview. *Applied Geography* 10: 247-264.

Beck, P.J. 1990b Regulating one of the last tourist frontiers: Antarctica. *Applied Geography* 10: 343-356.

Bergin, A. 1985 Recent developments in Australia's Antarctic policy. *Marine Policy* 9(3): 180-191.

Carrick, R. 1960 Conservation of nature in the Antarctic. *SCAR Bulletin* 6: 66-73.

Carrick, R. 1964 Problems of conservation in and around the Southern Ocean. In Carrick, R., M. Holdgate and J. Prevost, eds., *Biologic Antarctique; Antarctic Biology,* Hermann, Paris.

Chittleborough, G. 1984 Nature, extent and management of Antarctic living resources. In Harris, S., ed., *Australia's Antarctic Policy Operations,* Centre for Resource and Environmental Studies, Australian National University, Canberra pp. 135-161

Clark, M.R. and P. DINGWALL 1985, *Conservation of Islands in the Southern Ocean: Review of the Protected Areas of Insulantarctics. IUCN Commission on National Parks and Protected Areas, Gland.*

Codling, R.J. 1982 Sea-borne tourism in the Antarctic: an evaluation. *Polar Record* 21: 3-9.

Davis, B. 1984 Australia and Antarctica: aspects of policy process. In Harris, S., ed., *Australia's Antarctic Policy Options,* Centre for Resource and Environmental Studies, Australian National University, Canberra pp. 339-354.

Department of lands and survey 1983 *Management Plan for the Campbell Islands Nature Reserve.* Department of Lands and Survey, Wellington.

Department of lands and survey 1984 *Management Plan for the Snares Islands Nature Reserve.* Department of Lands and Survey, Wellington.

Department of lands and survey 1987 *Management Plan for the Auckland Islands Nature Reserve.* Department of Lands and Survey, Wellington.

Department of parks, wildlife and heritage 1991 *Macquarie Island Nature Reserve Management Plan.* Department of Parks, Wildlife and Heritage, Hobart.

Enzenbacher, D.J. 1992 Tourism in Antarctica: numbers and trends. *Polar Record,* 28 (164): 17-22.

Hall, C.M. 1990 *Tourism in Antarctica.* Institute of Australian Geographers Conference, University of New England, Armidale, September.

Hall, C.M. 1992a Tourism in Antarctica: activities, impacts, and management. *Journal of Travel Research*, Spring.

Hall, C.M. 1992b *Wastelands to World Heritage: Wilderness Preservation in Australia,* Melbourne University Press, Carlton.

Hall, C.M., S.Mcarthur, and P. Spoelder 1991 Ecotourism in Antarctica and adjacent Sub-Antarctic Islands: development, impacts, management and prospects for the future. In *Proceedings of the Ecotourism Conference*, Bureau of Tourism Research, Canberra.

Headland R.K. and P.L. KE AGUE 1985 Activities on King George Island group, South Shetland Islands, Antarctica *Polar Record* 22:475-484.

Hearder, J. 1989 The Australian Government perspective. In Handmer, J., ed., *Antarctica: policies and policy development.* Centre for Resource and Environmental Studies, Australian National University, Canberra, pp. 7-12.

Higham, T., ed. 1991 *New Zealand's Subantarctic Islands: A guidebook.* Department of Conservation, Wellington.

Holdgate, M.W. 1970 Conservation in the Antarctic, in Holdgate, M.W., ed., *Antarctic Ecology,* vol. 2, SCAR, Academic Press, New York, pp. 924-45.

Holdgate, M.W. and N.M. Wace 1961 The influence of man on the floras and faunas of southern islands. *Polar- Record* 10:475-93. House of Representatives Standing Committee on Environment, Recreation and the Arts [HRSCERA] 1989 *Tourism in Antarctica.* Australian Government Publishing Service, Canberra.

Josefa, S. 1990 Council against promoting tourism for Antarctica. *The Press*, July, p.16.

Keage, P.L. 1992 *The Conservation Status of Heard Island and the MacDonald Islands.* Occasional Paper No. 13, Centre for Environmental Studies, University of Tasmania, Hobart.

Keage, P.L. 1986 *Antarctic Protected Areas: Future Options.* Occasional Paper No. 19, Centre for Environmental Studies, University of Tasmania, Hobart.

Kriwoken, and L.P. Keage 1989a Introduction. In Handmer, J., ed., *Antarctica: Policies and Policy Development* . Centre for Resource and Environmental Studies, Australian National University, Canberra pp. 1-6.

Kriwoken, and L.P. Keage 1989b Antarctic environmental policies: protected areas. In Handmer, J., ed., *Antarctica: Policies and Policy Development.* Centre for Resource and Environmental Studies, Australian National University, Canberra, pp. 34-48.

Lyster, S. 1984 The Falkland Islands Foundation. *Oryx* 18:22-23.

Masson, D. 1990 Holidays to help the planet. *The Australian Magazine,* 3-4 March, pp. 50-52, 54, 56.

Nicholson, J. 1986 Antarctic tourism: The need fora legal regime? *Maritime Studies.* 29:1-7.

Peat, N. 1991 Treading softly in the sub-Antarctic. *Terra Nova* 5: 38-43.

Radio New Zealand 1992 Transcript of interview between Radio New Zealand 'Morning Report' presenter and Mr. Lew Sanson of the Department of Conservation, Invercargill, Tuesday, 26 May.

Reich, R.J. 1979 Tourism in the Antarctic: its present impact and future development. Unpublished thesis, Scott Polar Research Institute, Cambridge.

Reich, R.J. 1980 The development of Antarctic tourism. *Polar Record* 20: 203-214.

Roberts, B.B. 1977 Conservation in the Antarctic. *Philosophical Transactions of the Royal Society of London,* Series B, 279:97-104.

Smith, R.I.L. and H.W. Simpson 1987 Early nineteenth century sealers' refuges on Livingston Island, South Shetland Islands. *British Antarctic Survey Bulletin,* February, pp. 49-72.

Stephenson, L.H. 1990 Macquarie Island Nature Reserve - a philosophical approach to ecosystem conservation and the management of human impact. Unpublished thesis, University of Tasmania, Hobart.

The Australian 1990 Penguin stampede mystery, 25 June, 1990, p. 3.

Wace, N. 1990 Antarctica: a new tourist destination. *Applied Geography* 10: 327-342.

Appendix 1
Biogeographical Classification of Sub-Antarctic Islands

Classification	Islands	Description
Cool temperate	Tristan du Cunha Islands, Gough Island, He Amsterdam, He Saint-Paul, Antipodes Islands, Bounty Islands, Falklands off Tierra del Fuego.	Northern limit approximately the Subtropical Convergence, southern limit north of the Antarctic Convergence; trees and woody plants are common; mean annual temperature is generally higher than 5° C.
Subantarctic	Iles Kerguelen, lies Crozet, Heard Island, MacDonald Island, Marion Island, Prince Edward Island, South Georgia.	Islands in the vicinity of the Antarctic Convergence; phanerogamic communities dominate, with no trees; mean annual temperature ranges from 1° C to 5°C (with exception of warmer Iles Crozet.)
Mari time Anta retie	South Sandwich Islands, South Orkney Islands, South Shetland Islands, Bouvetoya.	Islands appreciably south of the Antarctic Convergence; cryptogamic flora, with no flowering plants; mean annual temperature is generally lower than 0° C.

Source: Clark and Dingwall, 1985, pp.186-87.

Index

Note: Figures are indicated by *italics*. Tables are indicated by **bold**. Endnotes are indicated by the page number followed by 'n' and the endnote number e.g., 20n1 refers to endnote 1 on page 20.